Specialized Injection Molding Techniques

Plastics Design Library

PLASTICS DESIGN LIBRARY (PDL)
PDL HANDBOOK SERIES
Series Editor: Sina Ebnesajjad, PhD (sina@FluoroConsultants.com)
President, FluoroConsultants Group, LLC
Chadds Ford, PA, USA
www.FluoroConsultants.com

The **PDL Handbook Series** is aimed at a wide range of engineers and other professionals working in the plastics industry, and related sectors using plastics and adhesives.

PDL is a series of data books, reference works and practical guides covering plastics engineering, applications, processing, and manufacturing, and applied aspects of polymer science, elastomers and adhesives.

Recent titles in the series

Biopolymers: Processing and Products, Michael Niaounakis (ISBN: 9780323266987)
Biopolymers: Reuse, Recycling, and Disposal, Michael Niaounakis (ISBN: 9781455731459)
Carbon Nanotube Reinforced Composites, Marcio Loos (ISBN: 9781455731954)
Extrusion, 2e, John Wagner & Eldridge Mount (ISBN: 9781437734812)
Fluoroplastics, Volume 1, 2e, Sina Ebnesajjad (ISBN: 9781455731992)
Handbook of Biopolymers and Biodegradable Plastics, Sina Ebnesajjad (ISBN: 9781455728343)
Handbook of Molded Part Shrinkage and Warpage, Jerry Fischer (ISBN: 9781455725977)
Handbook of Polymer Applications in Medicine and Medical Devices, Kayvon Modjarrad & Sina Ebnesajjad (ISBN: 9780323228053)
Handbook of Thermoplastic Elastomers, Jiri G Drobny (ISBN: 9780323221368)
Handbook of Thermoset Plastics, 2e, Hanna Dodiuk & Sidney Goodman (ISBN: 9781455731077)
High Performance Polymers, 2e, Johannes Karl Fink (ISBN: 9780323312226)
Introduction to Fluoropolymers, Sina Ebnesajjad (ISBN: 9781455774425)
Ionizing Radiation and Polymers, Jiri G Drobny (ISBN: 9781455778812)
Manufacturing Flexible Packaging, Thomas Dunn (ISBN: 9780323264365)
Plastic Films in Food Packaging, Sina Ebnesajjad (ISBN: 9781455731121)
Plastics in Medical Devices, 2e, Vinny Sastri (ISBN: 9781455732012)
Polylactic Acid, Rahmat et. al. (ISBN: 9781437744590)
Polyvinyl Fluoride, Sina Ebnesajjad (ISBN: 9781455778850)
Reactive Polymers, 2e, Johannes Karl Fink (ISBN: 9781455731497)
The Effect of Creep and Other Time Related Factors on Plastics and Elastomers, 3e, Laurence McKeen (ISBN: 9780323353137)
The Effect of Long Term Thermal Exposure on Plastics and Elastomers, Laurence McKeen (ISBN: 9780323221085)
The Effect of Sterilization on Plastics and Elastomers, 3e, Laurence McKeen (ISBN: 9781455725984)
The Effect of Temperature and Other Factors on Plastics and Elastomers, 3e, Laurence McKeen (ISBN: 9780323310161)
The Effect of UV Light and Weather on Plastics and Elastomers, 3e, Laurence McKeen (ISBN: 9781455728510)
Thermoforming of Single and Multilayer Laminates, Ali Ashter (ISBN: 9781455731725)
Thermoplastics and Thermoplastic Composites, 2e, Michel Biron (ISBN: 9781455778980)
Thermosets and Composites, 2e, Michel Biron (ISBN: 9781455731244)

To submit a new book proposal for the series, or place an order, please contact David Jackson, Acquisitions Editor
david.jackson@elsevier.com

Specialized Injection Molding Techniques

Edited by

Hans-Peter Heim

Universität Kassel, Institut für Werkstofftechnik,
Kunststofftechnik, Kassel, Germany

ELSEVIER

AMSTERDAM • BOSTON • HEIDELBERG • LONDON
NEW YORK • OXFORD • PARIS • SAN DIEGO
SAN FRANCISCO • SINGAPORE • SYDNEY • TOKYO
William Andrew is an imprint of Elsevier

William Andrew is an imprint of Elsevier
The Boulevard, Langford Lane, Kidlington, Oxford, OX5 1GB, UK
225 Wyman Street, Waltham, MA 02451, USA

Notices

Knowledge and best practice in this field are constantly changing. As new research and experience broaden our understanding, changes in research methods, professional practices, or medical treatment may become necessary.

Practitioners and researchers must always rely on their own experience and knowledge in evaluating and using any information, methods, compounds, or experiments described herein. In using such information or methods they should be mindful of their own safety and the safety of others, including parties for whom they have a professional responsibility.

To the fullest extent of the law, neither the Publisher nor the authors, contributors, or editors, assume any liability for any injury and/or damage to persons or property as a matter of products liability, negligence or otherwise, or from any use or operation of any methods, products, instructions, or ideas contained in the material herein.

ISBN: 978-0-323-34100-4

British Library Cataloguing-in-Publication Data
A catalogue record for this book is available from the British Library.

Library of Congress Cataloging-in-Publication Data
A catalog record for this book is available from the Library of Congress.

Publisher: Matthew Deans
Acquisition Editor: David Jackson
Editorial Project Manager: Peter Gane
Production Project Manager: Julie-Ann Stansfield
Designer: Maria Inês Cruz

Typeset by MPS Limited, Chennai, India
www.adi-mps.com

Contents

List of Contributors

Maik Feldmann Universität Kassel, Institut für Werkstofftechnik, Kunststofftechnik, Kassel, Germany

Johannes Fuchs Universität Kassel, Institut für Werkstofftechnik, Kunststofftechnik, Kassel, Germany

Tatjana Geminger Universität Kassel, Institut für Werkstofftechnik, Kunststofftechnik, Kassel, Germany

Hans-Peter Heim Universität Kassel, Institut für Werkstofftechnik, Kunststofftechnik, Kassel, Germany

Fabian Jakob Universität Kassel, Institut für Werkstofftechnik, Kunststofftechnik, Kassel, Germany

Stefan Jarka Universität Kassel, Institut für Werkstofftechnik, Kunststofftechnik, Kassel, Germany

Florian Mieth Universität Kassel, Institut für Werkstofftechnik, Kunststofftechnik, Kassel, Germany

Martin Rohleder Universität Kassel, Institut für Werkstofftechnik, Kunststofftechnik, Kassel, Germany

Mike Tromm Universität Kassel, Institut für Werkstofftechnik, Kunststofftechnik, Kassel, Germany

Preface

Dear Readers!

For many decades, injection molding has undisputedly been one of the most important methods for molding plastics. But why is that the case? What are the most noteworthy factors that have helped injection molding to be so successful? I believe that the following aspects are especially responsible for this success:

- *Plastic selection*: Almost every polymer will be able to be processed in the injection-molding process (duroplasts, elastomers, thermoplasts, biopolymers, highly filled materials, etc.).
- *Molding*: The manufacture of the final product takes place immediately. This means that in most cases, no or only minimal further processing of the manufactured injection-molded part is required after the manufacturing process of an injection-molded part.
- *Design*: Almost any type of design requirement, such as a complex component design, desired surface structures, or precise coloring, can be achieved in a process-integrated manner, and thus is inexpensive.
- *Automation*: The injection-molding process can be automated to a high degree by incorporating periphery devices. Therefore, it is suitable for the production of large quantities while simultaneously employing small cycle times per piece and guaranteeing a reproducible quality of the manufactured products.

These four aspects—plastic selection, molding, design, and automation—are fulfilled by every standard injection-molding process available on the market today. Additional options (i.e., specially customized plasticization units; automated tool changing units; highly efficient cooling systems; nanometer-precise, fine-structured tool surfaces; or special automation solutions for extremely short cycle times) are intended to increase efficiency further. As a consequence, a precise system concept before the initiation of series manufacture is one of the largest challenges for the individual responsible for the process, especially in the case of products with large quantities and long production periods.

In the early 1990s, an increased trend for function integration began, including the following:

- Decorative surfaces in the automotive industry
- Tools, household items, and sport items with haptic elements (soft components)
- Multicolor toys consisting of hard-soft combinations
- Foamed lightweight construction elements with a high surface quality for casings

These are only several examples, which the readers of this book are surely familiar with in one form or another.

All these requirements should fulfill the abovementioned aspects: plastic selection, molding, design, and automation, of course. In addition to their great efficiency, injection-molding processes should allow for function integration. Consequently, special injection-molding methods gained considerable significance and were substantially enhanced. New application fields and new solutions for special injection-molding methods resulted from the possibilities that this technology created. Depending on the counting method, more than 100 different injection molding methods are used today.

In 2002, Potente et al. [1] attempted to systematize the various methods. Figure 1 was modeled after this. Even though it is a reduced form, it still presents the essential special injection methods classified by their main characteristics (Figure 1).

Many special solutions, which were created several years ago, have almost completely disappeared from the market in the meanwhile. For instance, technologies like ultrasound injection molding or differential mode injection molding are hardly mentioned anymore. As another example, due to alternative manufacturing methods, the lost core technology used extensively for air-intake modules in cars in the 1990s has significantly lost market shares. In contrast, other methods have become more important; many current products would be unthinkable without composite technology and multicomponent injection molding.

Yet, every era has its trends. Currently, special materials are in fashion. Liquid silicon rubber (LSR) has significantly gained in importance, for example, and in the context of functionalization and the manufacture

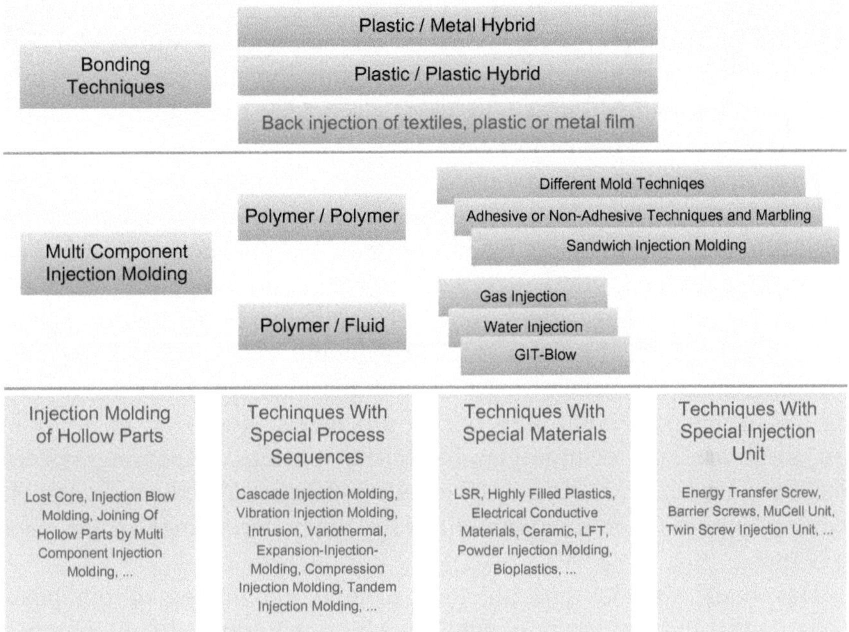

Figure 1 Schematic illustration of the special injection-molding methods classified according to their main characteristics.

of mechatronic systems, electrically and thermally conductive modified thermoplastic materials are of great interest. And also, owing to the notable improvement of the technical properties of materials based on renewable raw materials, the amount of bioplastics used in injection molding is increasing.

In this book, only bioplastics were focused on initially, and their processing behavior was elaborated. It becomes clear in the later chapters that they are generally suitable for all other special methods presented in this book so long as several special characteristics are taken into account. In the field of LSR, the research and development (R&D) activities appear to be increasing strongly today. A few sources (i.e., gas injection technology) have been available for several years, but a summary in the context of this book does not seem appropriate yet, because more new developments in terms of special injection methods of LSR are to be expected. This similarly applies to highly filled electric or thermally conductive functional materials.

Special solutions for very long flow paths, like cascade injection molding, extremely small components (microinjection molding), or the required facilities, are also not dealt with in this book. This is not because they are not important—it is actually the opposite. Their importance for plastics processing is beyond dispute, and several special solutions for these technologies are offered. But from the point of view of process-technological know-how these are not special injection molding techniques in the sense of this book.

What does this book focus on?

In addition to the injection molding of renewable raw materials, this book discusses the methods listed in Figure 1 in the categories of bonding techniques and multicomponent injection molding, for these two allocated groups have one thing in common: they require special process-technological knowledge. For example, this specific knowledge encompasses the interface stress of multiple components or local rheological and thermal conditions during processes. Moreover, in my opinion, these are key technologies for further plastics processing systems that have been customized by integrated, functionalized products. The required process combinations will have great importance in the years to come.

This book aims to provide the essential knowledge of composite technology and multicomponent injection molding in a practice-oriented manner. My co-authors and I hope that we achieved this well. Moreover, we also hope that you will be able to use this knowledge in your profession or education. Should this be the case, then we will be happy to have contributed in small part to your personal success, as well as to the success of plastics technology in general.

Hans-Peter Heim
Kassel, Germany
June 2015

Reference

[1] Potente H, Heim H-P, Ridder H, Kaiser E. It depends on the part – special injection moulding processes in automotive engineering. Kunststoffe Plast Europe 2002;92(3):10–13.

1 Multicomponent Technologies

Florian Mieth and Mike Tromm
Universität Kassel, Institut für Werkstofftechnik, Kunststofftechnik, Kassel, Germany

1.1 Introduction

In special injection molding processes (more precisely, in the case of multicomponent technologies), new processing alternatives continue to arise that enable material combination, and the economical manufacture of components with varying functionalities. Owing to its increasing significance in numerous areas and due to various motivations, multicomponent technology is often referred to as key technology. In the research literature, differing outlines, or classifications of the special processes and the according tool and handling technologies, can be found. Differentiations made in accordance with the process sequence are just as common as those made in accordance with the employed material classes or the required machine and handling technologies.

All these classifications have one thing in common—namely, their inability to provide fundamental insight into the processes that take place during the formation of the composite. This is a consequence of their specific methods of approach. Despite differing machine technologies and processing sequences that diverge from one another, the procedures during formation of the component are, for the most part, based upon identical basic mechanisms. Consequently, the formation of the composite can be influenced in a similar way. Therefore, each specific technical method must be analyzed only with regard to these processes, in order to provide information concerning the correlating processes that take place in the component and the possible effects that influence these processes.

We aim to provide the reader access to differential, in-depth information. Accordingly, this chapter has been structured into three sections. The first section "Basic Bonding Mechanism" is dedicated to theoretical information and describes the basic mechanisms.

Specialized Injection Molding Techniques.
DOI: http://dx.doi.org/10.1016/B978-0-323-34100-4.00001-8

Subsequently, in the second section "Influences of Material and Process Control on the Basic Bonding Mechanism", the focus shifts to interpretation and derives the effects of various material and processing parameters. Finally, the third section "Machines and Processes" closes with a pragmatic description of specific technical methods.

The chapter not only provides a fundamental knowledge base, but also enables the reader to understand methods that will become available in the future.

1.2 Motivation

Injection molded plastic-plastic composites are manufactured for very different purposes. The combination of properties and the integration of functions are the main motives. Components with different properties—color, hardness, flowability, thermal and electric properties, viscosity, etc.—are combined. Combining properties in this manner makes it possible to integrate functions specifically in the component or incorporate functional elements. For instance, in hard-soft composites, the hard component provides the strength while the soft component acts as a sealant or insulation, or performs haptic functions. With regard to optical or visual functions, various colors, transparent areas, or light-conducting components are combined with one another. Electric and magnetic components are employed for antistatic or shielding functions. In addition to functional surface modifications, decorative functions are utilized in back injection and back imprinting.

In most cases, the driving force behind combining plastics in an injection molding process is the aim to save costs by reducing the number of processing, handling, and assembly steps or joining processes.

In most applications, good adhesion between the components is desired. However, the opposite can also be the goal, in order to join movable elements by connecting them with nonadhesive or incompatible materials in one processing step, and, thus, reducing the number of assembly steps.

Many of these applications are consumer products (e.g., toothbrushes, toys, electronic devices, and kitchen utensils). Other examples can be found in medicine and automobile construction [1–5].

In general, the different processes or methods vary in the following ways:

- the material combinations used – compatible or incompatible materials;
- the actual combining of the plastics, which can take place in the plasticization cylinder, in the barrel in front of the screw, or in the mold;
- the chronological order of the joining of components – simultaneous or sequential; and
- the mold and handling technologies used.

Furthermore, depending on the process and method used, the components can either be in the same or different aggregate states at the time of joining.

1.3 Basic Bonding Mechanism

The adhesion between two components is generally defined by the adhesive strength (σ_H), which describes the resistance against separating loads, more precisely the inner strength (F_i) required to overcome the adhesion with regard to the real contact surface (A_W) [6]:

$$\sigma_H = F_i / A_W. \tag{1.1}$$

However, because these two factors can only be determined insufficiently or only with great effort in practice, composite strength (σ_V) is used as an evaluation parameter. It is defined by the outer strength (F_a) with regard to the initial cross section (A_N) [6,7]:

$$\sigma_V = F_a / A_N. \tag{1.2}$$

Thus, the composite strength defines the mechanical strength and conditionally or indirectly correlates with the actual adhesive strength, which is based upon adhesion phenomena [6].

The formation of the composite and the according composite adhesion of two components are both determined by the overlapping of various mechanisms and influencing factors, which in turn are essentially caused by the material properties, the processing method, and constructive factors. In most cases, these mechanisms and their reciprocal effects have not yet been fully understood [2]. Too little is known about bonding mechanisms that occur in real life, and, above all, which proportion of the adhesion they contribute to [1]. The technological and constructive influences on the process are the locally differing flow and temperature conditions, which define the orientation conditions in the boundary layer of the composite partners. In contrast, the material properties determine the wetting behavior, the surface tension, and the adhesion properties. In plastic-plastic composites, an overlapping of material and process-dependent mechanism in correlation with constructive boundary conditions can be assumed [6].

When injection molding thermoplastic plastic-plastic composites, chemical primary valence bonds are unlikely [6,8]. The secondary valence forces that take effect in the various adhesion theories (i.e., van der Waals forces, dispersion and dipolar forces, and hydrogen bonds) can create effects in only a very limited range of tenths of a nanometer [6, p. 23]. For this reason, wettability and sufficient convergence between the composite partners are prerequisites for good composite formation. Principally, the formation of the interfacial layer between the joining partners can be compared to the formation of a joint line [9]. As is also the case when welding plastics, diffusion processes play a role. Sufficient molecular mobility and diffusion processes across the interfacial layer are necessary for these actions to take place [9]. Nevertheless, it remains unclear whether the diffusion speed is high enough to play an essential role in technical processes that are completed at a high processing speed.

The bonding mechanism in general takes place as a result of adhesion and cohesion phenomena. Adhesion in the context of plastic-plastic composites describes the bonding forces between two materials and is described in the following section. Cohesion describes the forces of attraction within a material, meaning the bond between atoms and molecules created by mutual attraction [1]. The effect of cohesive bonding mechanisms on plastic-plastic composites is not yet clear. According to Ref. [8], cohesive forces are not regarded as significant in correlation with the formation of the composite.

1.3.1 Adhesion

Adhesion is defined as the connection between two materials in contact with one another. Various adhesion theories exist that can contribute to composite formation and the required mechanisms. However, no single theory explains all ongoing processes. Adhesion, therefore, can be considered the sum of various simultaneously occurring bonding forces, which are all based on different adhesion theories.

Adhesion is usually classed as one of two varieties: mechanical and specific adhesion [4,8]. Yet, other categorizations are common (i.e., chemical, physical, and mechanical adhesion) in the literature [e.g., 10]. Mechanical adhesion describes the intrusion of a component into the pores and depressions of the other component, respectively. Thus, this form of adhesion is essentially based upon the surface roughness and surface structures of the components. Specific adhesion is divided into chemical, physical, and thermodynamic reciprocal effects. Moreover, it is classified into the following theoretical approaches:

- Chemisorption—This theory is based on the formation of chemical bonds in the interface, but, according to Ref. [8], has not yet been verified for plastic-plastic composites.
- Polarization—This theory (by De Bruyne) states that owing to polarity, molecular physical interactions between the materials result in adhesion processes. Here, sufficient wetting is also a requirement. This is the case if the polarities of both components are identical [4,11]. When combining polar and nonpolar plastics, adhesion as described in the polarization theory is not possible [6]. Functional groups situated on the surface (carboxyl, hydroxyl, and carbonyl groups) can contribute to improved composite adhesion [8]. The theory supplies an essential contribution to adhesion for plastic-plastic-composites.
- Electrostatic—This theory (by Derjaguin) focuses on the formation of an electric double layer in the interface that has been induced by the potential differences between the components [8]. The hypothesis states that the more similar the materials are, the smaller the double layer will be, and thus the adhesive strength as well [8]. This theory has been criticized because it cannot explain the adhesion between similar components [6]. The electrostatic theory is not being followed up in current research with regard to plastic-plastic composites [8].
- Thermodynamic—This theory is based upon the equation by Dupré, which defines the work of adhesion for separating two surfaces in correlation of interface and surface stresses. Good adhesive strength requires an interface stress

that is close to zero, which occurs when the polar and dispersed proportions of the according surface stresses are equal [6]. Owing to the very limited range of intermolecular forces (van der Waals forces, hydrogen bonds, and primary valence forces), sufficient closeness of the molecular chains or sufficient contact between the components is a fundamental requirement [4,11]. With regard to plastic-plastic composites no clear statement is given in the literature.

- Weak boundary layer theory—This theory assumes the formation of an intermediate layer in the interface that counteracts direct contact between the joining partners or composite materials. Reasons for this occurrence can be impurities in the surface, air or gas pockets due to poor wetting, or products of reactions or substrates in the surrounding air. This theory is often used when adhesion phenomena cannot be explained by other theories [8].
- Diffusion—In literature, the diffusion theory is often referred to as part of specific adhesion, and is based upon different models. Due to its high significance concerning composite formation in plastic-plastic composites, it will be described in depth in Section 1.3.2.

1.3.2 Diffusion

The term diffusion generally refers to the autonomous equalization of the concentration in the form of directed molecular movement. The driving force is the difference in concentration between different phases.

With regard to plastic-plastic composites, the diffusion theory implies a diffusion of the molecules beyond the interface of the contacting elements. On a molecular level, mutual diffusion of molecular chains close to the surface or mutual penetration into the surface layer of the other component takes place.

After diffusion has completed, there is no clear interface between the composite partners. A boundary layer with molecule segments of both components is available. The interface consists of a mixture of molecules or molecular segments of both components. Moreover, the molecule chains are intertwined with one another. The entanglement of the molecules results in good composite strength [7]. Diffusion is a thermodynamic process, not a phenomenon that only takes place on the surface

[12]. The prerequisite for diffusion is contact between molecules [1]. Diffusion processes are possible if a thermodynamic compatibility of the components and a sufficient amount of molecular mobility are available. A temperature (T) higher than crystallite melting temperature (T_C) – $T > T_C$ applies for semicrystalline plastics, and a temperature (T) much higher than glass transition temperature (T_G) – $T \gg T_G$ is applicable for amorphous plastics. According to Ref. [6], diffusion processes were also observed in incompatible systems, but only for small ranges [53, p.381]. This can be explained by the partial diffusion of macromolecular segments [6].

Diffusion theory largely contributes to explaining composite strength. Sufficient molecular mobility is a prerequisite because diffusion speed depends on the mobility of the molecule or molecular segment [1], which is essentially determined by the interface temperature and the molecular weight [6]. The mathematical description of diffusion processes is Fick's law of diffusion; however, it is not directly applicable to diffusion processes in polymer chains [4,10]. The diffusion theory basically makes it possible to describe autoadhesion between two plastics. Yet, the theory displays deviations if semicrystalline plastics, strong chemical reciprocal effects, or both are present [13]. It is assumed that Fick's law of diffusion is applicable only if the thickness of the boundary or contact layer reaches the thickness of a bundle of molecules. Prior to diffusion, the chain form has a special significance. Entangled chains inhibit diffusion, and require more energy to accomplish the diffusion process, because they are intertwined with other chains and are restricted in terms of their mobility. In the moments before the surface layer width is achieved—as mentioned previously—the diffusion can be described using the reptation model [4,10]. The penetration depth is described using Einstein's Law (as described in the next section).

The transition in the area of the interface, which is caused by diffusion processes and is not clearly defined, leads to a reduction of the property differences of the components in this section. This is generally advantageous for the composite strength. Longer contact times, higher temperatures, a higher degree of chain flexibility, the level of branching, a low molecular weight [2], and a small discrepancy between the solubility parameters [12] all prove to be beneficial in diffusion processes. Factors that limit the mobility of the molecular chains (i.e., high levels of branching, high degrees of crystallinity, strong polar groups, and a high content of filling material) negatively affect diffusion [12].

In general, diffusion is temperature-dependent and correlates with the viscosity, meaning that it depends upon the type of process and process control. With regard to conventional joining processes such as welding, the diffusion periods are often too long, compared with processing time, to be able to explain the achievable composite strengths [7,10]. By injection molding plastic-plastic composites, the actual time in which high temperatures are used to achieve sufficient chain mobility is also normally quite small because of the typically low wall thickness and high cooling rates. Thus, the timeframe for diffusion processes can be expected to be very low.

1.3.2.1 Diffusion Theory—The Einstein Equation

The diffusion theory describes transposition processes, which occur when two polymers come into contact. The formation of the composite is influenced by: the number of molecular chains that have penetrated each other, the penetration time, and the types and sizes of the forces between the molecules. The hypothesis in this model employs a simplified perspective, stating that mainly entirely stretched, free chain ends take part in the diffusion process. This assumption enables an estimation of the diffusion coefficient. Einstein's Law describes the intermediate penetration depth $\tilde{x}$ of the diffused chains in the diffusion partners in correlation with the diffusion coefficient (D) and time (t). The penetration depth increases with the square root of the contact time [1,10]:

$$\Delta\tilde{x} = \sqrt{(2^*D^*t)} \tag{1.3}$$

The penetration depth and diffusion layer thickness depend upon the compatibility of the polymers. In the case of compatible systems, this layer can equal several micrometers [1]; in partially compatible systems, several ångström, or up to several nanometer [12].

After a certain period of time, the number of diffused molecules will no longer change [4]. The deeper the depth of penetration and the higher the number of diffused chain segments, the better composite formation will be.

The number of diffused molecules can be determined at a specific point in time to estimate the adhesive strength. For this purpose, an intermediate penetration depth of an entirely stretched molecule and knowledge of the intermediate deformation speed are needed. The precise basis for calculations can be found in literature [e.g., 10].

1.3.2.2 The Reptation Model

Fick's law of diffusion describes the flow of the material along a concentration gradient in diffusion processes in the dependency of the diffusion coefficient—this describes the tendency to diffuse and the diffusion speed. However, it cannot be directly applied to the diffusion of polymer chains, especially not for entangled polymer chains.

It is assumed that Fick's law of diffusion is initiated when the surface-layer width reaches the thickness of a molecule bundle. In the time preceding this moment, the chain form is of high significance. Molecular chains are larger and heavier than single molecules; thus, their tendency to diffuse decreases the longer they are and the higher their degree of entanglement is. Furthermore, they require more energy for diffusion processes. In regard to a good composite formation, entanglements with the molecular chains of the bonding partner are particularly advantageous.

The reptation model (according to De Gennes and Doi and Edwards) observes molecular movements in the melt at various points in time. Temperature-dependent reptation time refers to the time after the molecular chain has left its original orientation. In most cases, free molecular chains take a starkly tangled form. Their movement is described as snakelike or wormlike. Owing to the tangled structure of neighboring molecules, the chains are trapped in a form of a network. This is often illustrated with a pipe or hose that is surrounded by molecular chains, and it aims to represent the molecular chains' limited freedom of movement (Figure 1.1).

Two main movements are possible in the "pipe." The chain ends can move freely, thus changing the form of the chain and eventually leaving the original "pipe" position. After an initial, heat-induced stretching process (Brownian molecular movement), the chain slowly but surely leaves its original orientation and, by the end of the reptation, has completely left its primary form. The reptation time can be calculated using the Arrhenius approach. By means of the reptation time and an intermediate penetration depth, estimations concerning the composite strength can be made. For further information, please refer to the relevant literature [4,10].

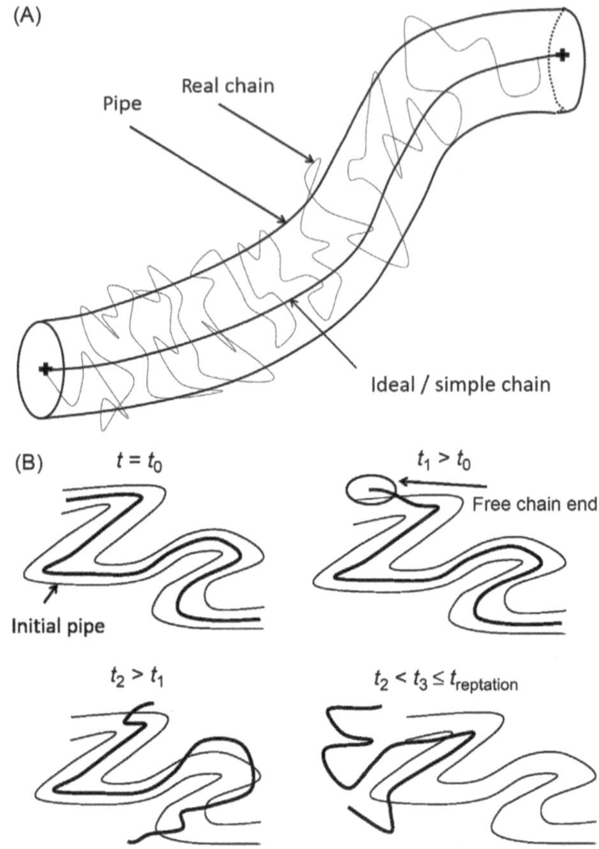

Figure 1.1 Reptation model according to Ref. [10]; compare with Ref. [14].

1.4 Influences of Material and Process Control on the Basic Bonding Mechanism

When injection molding plastic-plastic composites, adhesion is a function of the material, process control, and part design. These factors influence each other and define the local rheological and thermodynamic conditions [7]. As previously mentioned in the introduction to this chapter, all processing parameters, material effects, and basic conditions either directly or indirectly affect the interface, and thus the quality of the composite as well. In the following section, the influencing factors of the bonding determinants relevant in practice will be allocated to three bonding determinants: interface compatibility, interface temperature, and interface stress (Figure 1.2).

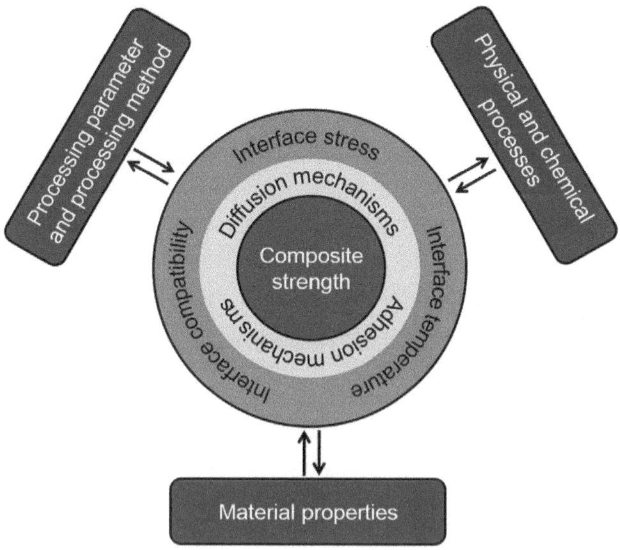

Figure 1.2 Influence factors of the bonding determinants that are relevant in practice.

1.4.1 Interface Compatibility

The term interface compatibility refers to the extent to which the material properties of composite partners promote, complicate, or prevent composite formation in the interface. The properties result mostly from a complex interaction of processing conditions, and thus they are rarely constant.

1.4.1.1 Surface and Interface Tension

The adhesion between two components is described by the Dupré equation, and depends upon the surface and interface tensions [2,12]. The equilation describes the work needed to seperate the components from one another.

The term surface tension refers to forces that take effect on the molecular level and aim to keep the surface of a liquid small. The surface energy specifies the amount of energy needed to break the chemical bonds when creating a new surface. In liquids, the surface tension equals the surface energy. In solid materials, this value can be measured only indirectly. The energy, or tension, determines the wettability and adhesion ability and is defined by the chemical structure of the materials [7]. It is based on the intermolecular interactions and various types of bonds [2]. The surface tension consists of a nonpolar, disperse portion and a polar portion [2]. The composite strength depends on both portions [7]. The lower the difference between the surface tensions of both components meaning the closer the surface tension ratio is to 1, the better the composite strength [15]. However, there are exceptions—i.e., the insufficient bond strength of polyolefin combinations or PA6-PMMA combinations that can be explained by shrinkage discrepancies [1]. The polar portion of the surface tension can be increased by surface treatment. If the tension ratios strongly diverge, zero adhesion can occur. A prime example of where zero adhesion is specifically employed is assembly injection molding (see Section 1.5.1) [12].

The surface tension is also influenced by the process parameters, such as the temperature and the process-induced degree of crystallinity. As the temperature increases, the surface tension reduces. Thus, during processing (at high temperatures), lower stress values are on hand in a material than for room temperature [1]. Usually, the reduction of surface tension does not occur equally in both components, thus the ratio of the surface tension between the materials also differs during processing [1]. According to Ref. [1], the following equation enables an estimation of the surface tension ($\sigma_{(T)}$), in regards of the contact temperature. As a requirement the surface tension at start temperature ($\sigma_{(To)}$) and the specific volume of the polymer at start temperature ($v_{(To)}$) have to be known. The specific volume can be found in the p,v,T – diagram for the polymer. Furthermore the molecular weight of the polymer is assumed to be constant.

$$\sigma_{(T)} = \sigma_{(To)} * \left(\frac{v_{(To)}}{v_{(T)}}\right)^4. \tag{1.4}$$

Generally, crystalline phases display a higher surface tension than amorphous phases [1]. In plastic-plastic composites, for identical components, this can lead to varying values because the second component is in a melted (and thus an amorphous) state at the time of the contact. During the injection molding process, in this context, it is important to take into account that the first component possesses an amorphous surface layer owing to the rapid cooling conditions in these areas [1].

The term interface tension denotes the forces between two phases that are in contact, and is essentially defined by the surface tensions of the single phases. With regard to plastic-plastic composites, the lower the interface tension, the more adhesion and diffusion processes will be promoted [7,12]. When the same components are in contact, the interface tension is zero [12]. The interface tension is indirectly influenced by the surface tension, due to both material and processing parameters. Thus, additives, the molecular weight, the degree of crystallinity, and processing temperatures affect the interface tension [12].

1.4.1.2 Crystallinity

The crystallinity of a material strongly affects the molecular mobility, and, in the case of semicrystalline materials, has an essential influence on the composite strength. In extreme cases, this can lead to disadhesion, even in compatible materials [9]. The molecular chain mobility is limited in crystalline areas. As stated previously, a high molecular mobility facilitates diffusion processes. An increasing degree of crystallinity, whether material- or process-induced, counteracts the diffusion tendency, and, thus also the strength of the composite [16].

When injection molding semicrystalline plastics, a surface layer with a low degree of crystallinity, small crystallization nuclei, and spherulites is developed due to the rapid cooling of the mold. Reheating initiated by contact with the second component can cause the crystallization nuclei and spherulites to function as crystal nucleus and can result in a postcrystallization of the surface area. Similar effects have been observed for a later tempering process [6]. Should reheating occur, diffusion processes can take place if enough thermal input is available. This, in turn, promotes composite strength. If the thermal input from the second component is insufficient to melt the contact area, this can result in poor composite strength. This is due to diffusion processes being limited in the melt of the second component because it is injected beside postcrystallized areas and spherulites [12].

1.4.1.3 Molecular Weight

The influence of the molecular weight is characterized by two opposing effects. On the one hand, as the molecular weight increases, the molecular chain mobility decreases, which is particularly counterproductive for diffusion processes. On the other hand, the possibility to diffuse entire molecule tangles increases. The resulting, higher density of entanglements leads to increased composite strength. The optimal composite strength correlates with the molecular weight [6,12]. In general, a critical, minimum molecular weight exists. Below this minimum, the chains are too short for sufficient diffusion and entanglement processes, and, therefore, they cannot adequately contribute to the formation of the composite. The critical weight varies depending on the polymer [6]. The correlation of molecular weight and composite adhesion is displayed qualitatively in Figure 1.3.

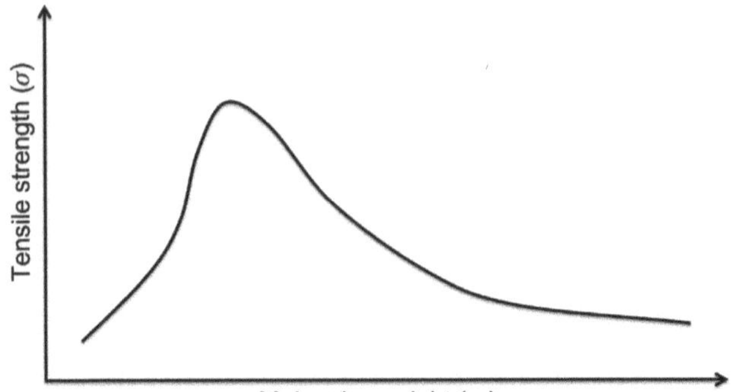

Figure 1.3 Qualitative influence of the molecular weight on the composite strength according to Ref. [12].

1.4.1.4 Polarity

The polarity has a decisive influence on the wetting. Good wetting is given if the polarities of the two components are identical [4,17]. Adhesion is not realizable if nonpolar and polar plastics are combined [6]. Functional groups at the surface (carboxyl, hydroxyl, and carbonyl groups) can help to achieve improved composite adhesion [8].

High polarity generally induces lower molecular mobility and thus inhibits diffusion processes. This, in turn, has a negative effect on the composite strength.

1.4.1.5 Molecular Orientations

The orientation of the molecules in the surface layer significantly influences the interface-transcending relaxation and diffusion abilities [4]. The orientation depends on the position in the flow cross section. Molecules in the outer layers are oriented parallel to the flow direction, molecules in the middle of the flow channel are less orientated in direction of the flow (Figure 1.4A). With regard to the composite formation, molecules that are oriented parallel to the interface can participate in the diffusion process at high temperatures, and, in doing so, contribute to enhancing adhesion. The higher the degree of orientation in the molecular chains, the larger the potential for elastic back deformation of the molecule, and, thus, the ability of chain segments to diffuse beyond the boundary layer. These correlations are shown in Figure 1.4B and C.

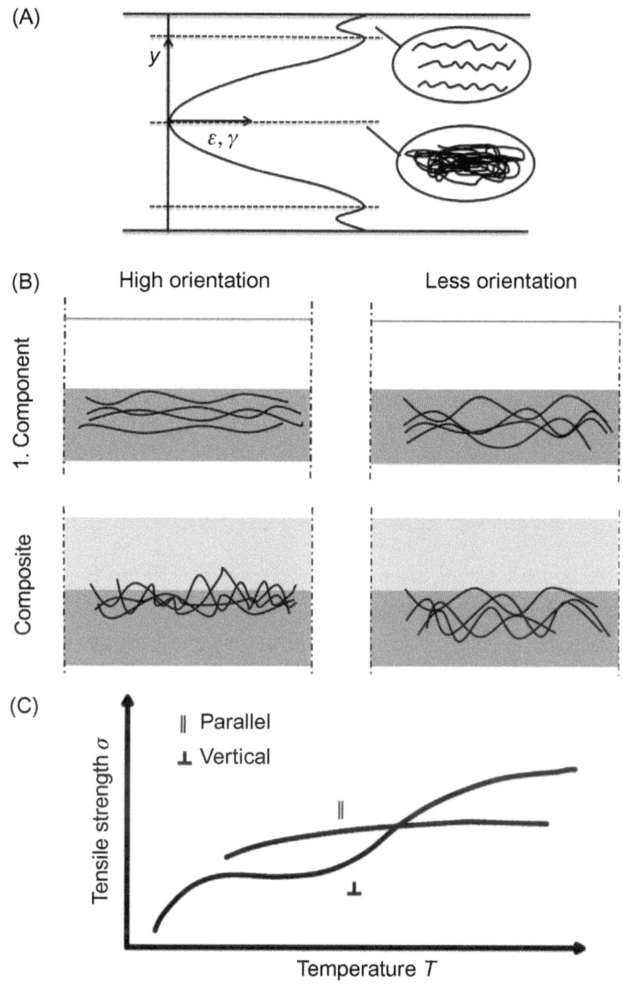

Figure 1.4 (A) Formation of orientations in the injection molding process, according to Ref. [4]; (B) Significance of orientations for diffusion, according to Ref. [4]; (C) Effects of orientations on the composite strength, according to Ref. [12].

1.4.1.6 Thermodynamic Compatibility/Miscibility

Thermodynamic compatibility or miscibility refers to the mutual penetration of molecules during diffusion processes [7]. The molecular mixing process is determined by thermodynamic processes (mixing enthalpy and mixing entropy) and influences the interface layer thickness.

Semicrystalline plastics are, in most cases, not capable of crystallizing together (isomorphic), making mixing only possible in the amorphous areas. Here, the miscibility can be estimated in limited areas by means of the glass transition temperature. If complete miscibility is available, only one glass transition temperature will be measurable. However, if the materials are not compatible, differing glass transition temperatures of the individual components occur. [6,7]. Good miscibility promotes diffusion processes and positively affects the composite strength.

The solubility parameter describes miscibility as a prerequisite for diffusion processes. It is the square root of the cohesion energy density and is defined by secondary valence forces. If the parameters of both components range close to one another, a low mixing enthalpy is available and positively influences the miscibility and the composite strength. Strong, polar reciprocal effects (i.e., hydrogen bonds) complicate or prevent an accurate assessment of the mixing behavior by means of the solubility parameters. Surface pretreatment processes can also influence the solubility parameters [6,8,13].

1.4.1.7 Surface Pretreatment

The aim of a surface pretreatment is a fine cleaning or activation of the contact area to influence the surface and interface tensions. This should lead to adhesion improvement, with the purpose of increasing composite strength. Most often, it is used for material combinations with medium to zero adhesion, or to level out the effects of bad processing conditions. Owing to the additional means to influence the adhesion properties, processes with low temperatures or pressures can be used without losing adhesion. The most commonly used methods are adhesion modifications using primers, flaming, or corona/plasma methods. All these methods can be integrated into an injection molding process in normal, atmospheric conditions via handling systems or rotary disks [2]. For more information, please refer to the according literature [8,17–24].

1.4.2 Interface Temperature

The interface temperature and its chronological development are commonly viewed as the two most essential parameters of influence referred to composite formation by plastic-plastic composites. The temperature that results upon contact of both components fundamentally determines the formation of a boundary layer. The longer a higher interface temperature persists, the higher the degree of molecular mobility, and the higher the according tendency for diffusion processes will occur.

1.4.2.1 Mass Temperature/Injection Sequence

The required minimum interface temperature limit for a composite formation depends on the material. For example for polypropylene (PP), this temperature is defined at 130 °C. But for maximum composite strengths higher temperatures in the interface are necessary [9]. The melt temperature of the first component is, as a rule, unimportant, because often the first component has to be cooled down intensively to achieve stability in the molded part or dimensional precision. Insufficient cooling of the first component can lead to a deformation of the contact area due to the melt pressure of the second component [1]. If the first component is cooled down too much, zero adhesion can even occur in compatible materials if the heat input of the second component is too low [12].

The mass temperature of the second component should be selected in regards of its ability to melt the boundary layer of the other component. But, if the temperature selected is too high, for example by using another material, too much melting may occur, and cause deformation in the first component.

A contact temperature higher than glass transition temperature is recommended for a longer period of time for amorphous plastics. A contact temperature higher than crystallite melting temperature is suggested for semi-crystalline plastics. These temperature ranges enable a macro-Brownian mobility of the molecular chains, which is helpful for diffusion processes [9]. In regards of the thickness of the layer that needs to be melted to achieve good composite adhesion, literature suggests values between 10 nm and 10 μm [1,2]. This wide range can be explained by the fact that the amount of influence strongly depends on the materials and the material combination [6,9,12,13].

In composite injection molding, the interface temperature is influenced by several parameters. In addition to the temperatures of the components, the thermal diffusivity plays an essential role as well. It is defined by the density, the specific thermal capacity, and the thermal conductivity. The degree to which the melting temperature of the second component can be influenced is restricted, because it is material dependent, and cannot be varied to whichever desired degree. If the first component consists of a semi-crystalline plastic with a temperature that is lower than the crystallite melting temperature (T_C), a high amount of energy is needed to re-melt the crystalline areas. As a consequence, less energy is available to increase the interface temperature. This correlation can be seen in Figure 1.5.

Figure 1.6 provides an exemplary illustration of the temperature conditions in the interface of a butt joint. Here, two different interface

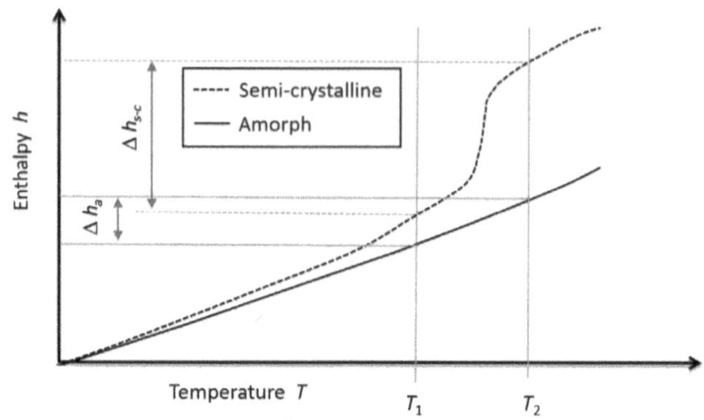

Figure 1.5 Qualitative trend of enthalpy by temperature for amorphous and semi-crystalline materials—for semi-crystalline materials applies: $T_1 < T_C$ and $T_2 > T_C$.

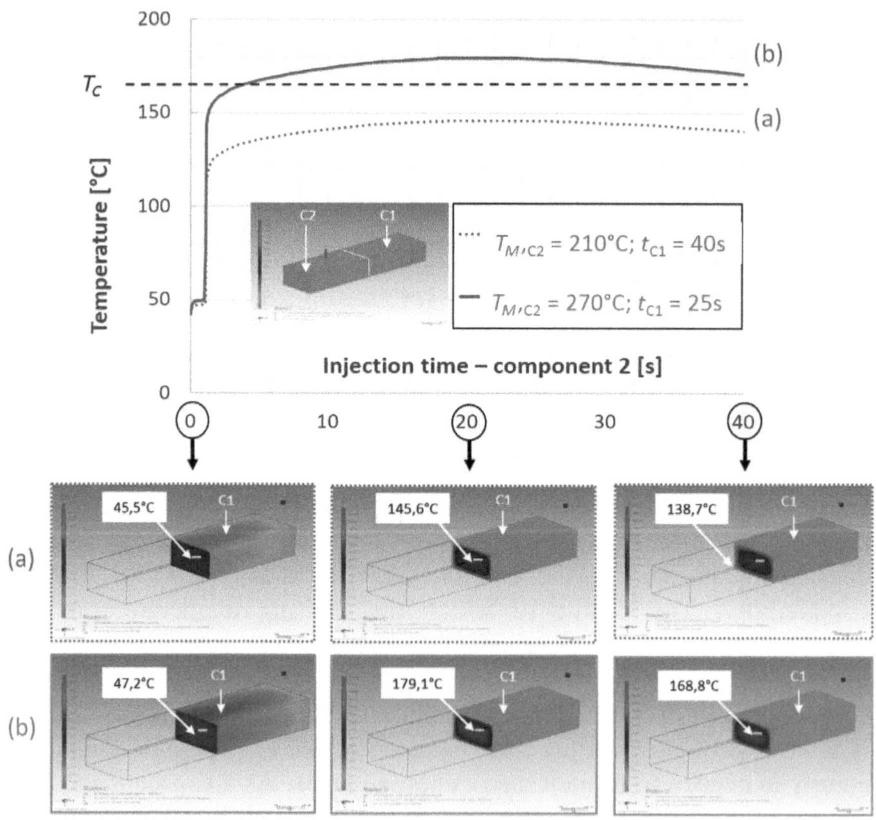

Figure 1.6 Temperature conditions in the interface for a combination of PP–PP; (a) melting temperature component 2 = low / cooling time component 1 = high; (b) melting temperature component 2 = high / cooling time component 1 = low.

temperatures are shown, resulting from two combinations of melting temperature of the second component and cooling time of the first component. Both components are of PP; the second component is directly injected after the end of cooling time (EOF) of the first component. The melt temperature of the first component was constant for both cases. In case (a) the melting temperature of the second component ($T_{M,C2}$) was set to 210 °C, which is the minimum suggested melting temperature for PP. The cooling time of the first component (t_{C1}) was set to 40 s. In case (b) the melting temperature of the second component ($T_{M,C2}$) was set to 270 °C, which is the maximum suggested melting temperature for PP. The cooling time of the first component (t_{C1}) was set to 25 s, which is the suggested cooling time for this part.

As can be seen in the graph, the resulting temperature in the interface of both components is for case (b) at a higher level and for case (a) at a lower level than crystallite melting temperature, for a long period of time. As also mentioned, the minimum required interface temperature of 130 °C for PP for composite formation is given for both cases, but for case (b) higher composite strength is expected.

The temperature in the interface is also influenced by the type of interface design. Figures 1.7 and 1.8 show the two, essentially different cases of a butt joint (1.7) and an overlapping joint (1.8).

In general the heat dissipation in the interface is lower than in direction to the mold wall.

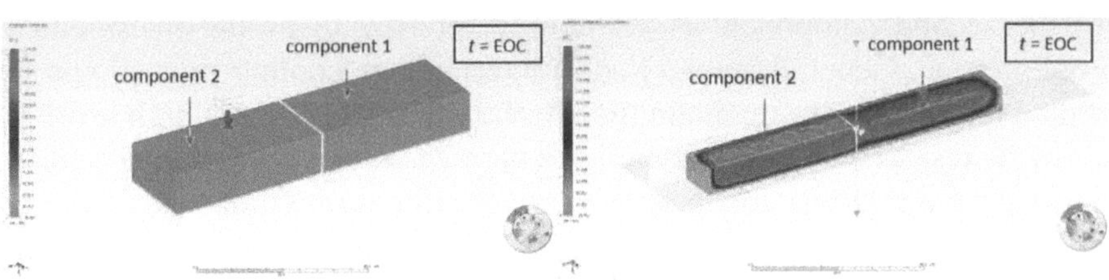

Figure 1.7 Temperature distribution in the interface of a butt joint connection—time in process = end of cooling (EOC)/ejection.

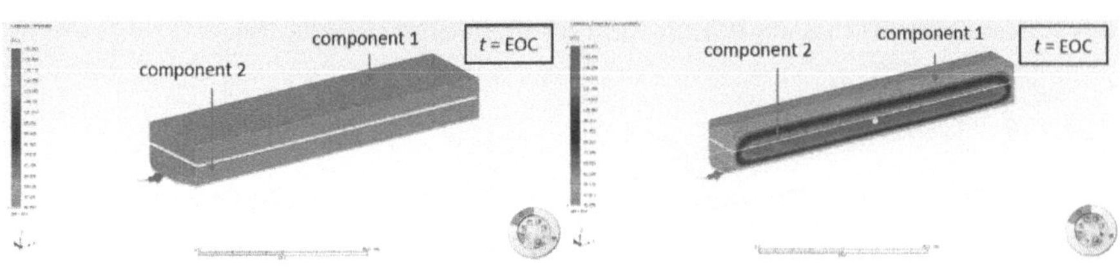

Figure 1.8 Temperature distribution in the interface of an overlapping connection—time in process = end of cooling (EOC)/ejection.

The heat dissipation in the interface is much lower than to the cooled mold wall. As a consequence, the temperature in the interface remains at a higher level for a longer period of time. Regarding the cross-sections, the temperature maximum is displaced toward the interface by the second component. For an overlapping joint, often, the interface area is larger, than for butt joints, thus more time at higher temperatures is on hand in the interface area.

1.4.2.2 Mold Temperature

The mold temperature significantly influences the interface temperature, crystallization, and shrinkage effects. It can have either positive or negative effects [2].

Generally, a high mold temperature results in improved wettability of the first component and a higher temperature level in the surface layer. Thus, it often leads to higher composite strength [1,13]. However, even low mold temperatures can promote diffusion processes for semicrystalline materials. The reason is, that for high cooling rates, less crystallinity in the surface layers is created [7].

1.4.2.3 Intermediate Cooling Time

In composite injection molding, the period of time between end of injection and holding pressure phase of the first component until injection of the second component is commonly referred to as the intermediate cooling time. Generally, a very short intermediate cooling time should be aimed for, in order to maintain the interface temperature at a high level.

In the research literature, there have been cases in which an intermediate cooling time that was too short contributed to poorer composite strengths and high distributions in strength. There is a time limit at which the highest composite strength can be achieved. Indirect factors of influence are suspected to be the reason—namely, the influence of shrinkage (in particular, short-term shrinkage), shrinkage-induced altered heat transfer, and crystallization processes which lead to shrinkage effects [9].

1.4.2.4 Contact Time

Provided that a sufficient joining force is available, a time limit exists in which the composite formation can be influenced positively. Above this time limit higher contact times show no additional effects. A higher thermal input leads to an accelerated diffusion, and, thus, can lead to comparable strengths within shorter times and increased achievable strength values [6,12]. The qualitative correlation can be seen in Figure 1.9.

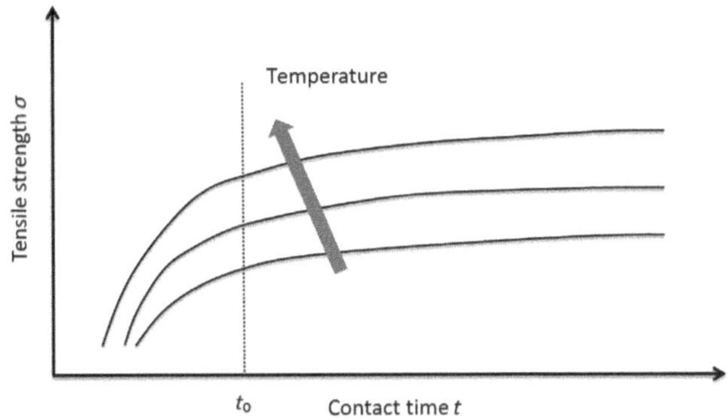

Figure 1.9 Composite strength in correlation with the contact time and the interface temperature according to Ref. [12].

1.4.3 Interface Stresses

Tensions in the interface can be triggered by countless factors. They can be process induced, or, at a later point in time, they can be triggered into stress states by various properties of the joined components when in use (i.e., heat expansion, differences in E moduli, and different shrinkage values). Tensions can negatively influence the composite strength to a large extent.

1.4.3.1 Shrinkage

Shrinkage is caused by processing of material. There are three types: demolding shrinkage, processing shrinkage, and postshrinkage. Demolding shrinkage refers to molded part deviations that occur directly after the demolding process. Processing shrinkage is defined as the shrinkage effect that takes place 16 h after manufacture. All later shrinkage effects (i.e., relaxation of internal stresses, reorientation processes, and postcrystallization) are called postshrinkage, and are not categorized over a specific time period (Figure 1.10). Shrinkage can be influenced by numerous processing parameters, especially in the case of semicrystalline plastics. In plastic-plastic composites, all shrinkage proportions can take effect depending on the type of process control.

Differing shrinkage behavior of various components can lead to stresses in the interface. Here, shrinkage that takes place perpendicular to the connection area can be critical and negative in regard to the composite strength [7]. However, process-induced and geometry-induced stresses can also occur in identical components. For instance, the higher temperature of the second component can lead to different processing shrinkages upon coming into contact with the much cooler first component, thus, creating stresses.

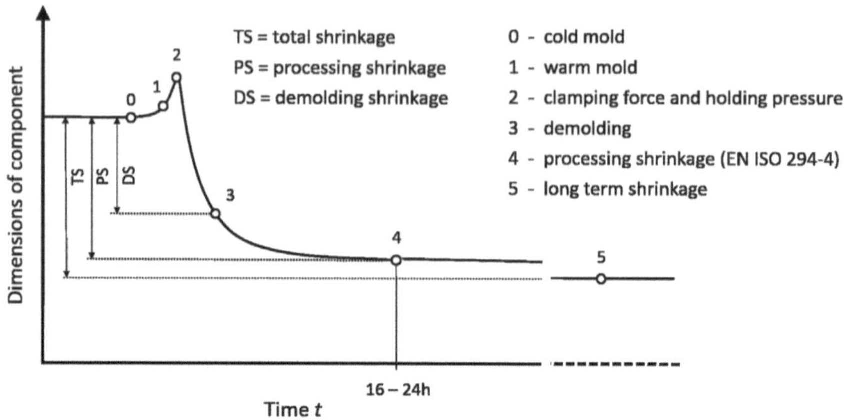

Figure 1.10 Chronological sequence of shrinkage, according to [9].

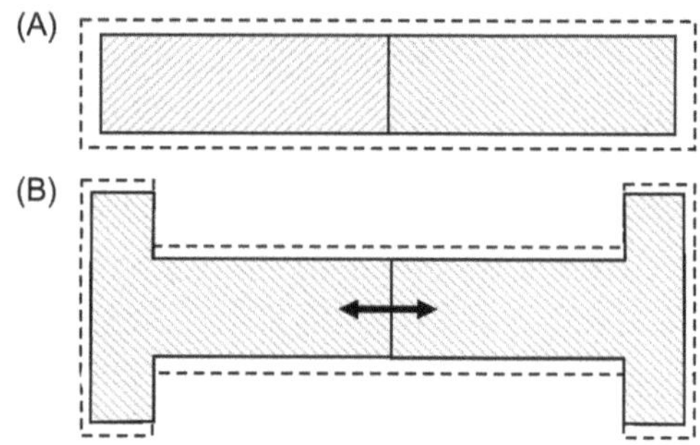

Figure 1.11 Formation of shrinkage in butt joints and overlapping joints: (A) free shrinkage and (B) form bound shrinkage.

At the moment that the melt of the second component comes into contact with the first component, the shrinkage phenomena begins and influences composite formation. The geometry of the joining zone and of the component itself has a major influence in this context. In this regard, a differentiation is made between free shrinkage and form bound shrinkage. Components can shrink in all directions when free shrinkage is taking place. Form bound shrinkage refers to the restriction in the shape of the cavity, which is predefined by the sprue, undercut, ribs, and connecting parts, and influences the shrinkage. If the direction of shrinkage leads away from the interface, the form bound shrinkage can initiate higher tensions in the interface, and, therefore, it can cause poorer composite adhesion [7]. The influence of the shrinkage in dependency of the type of joint can be seen in Figure 1.11.

1.4.3.2 Connection Design/Interface Geometry

The type of connection design offers many possibilities for influencing the composite strength. Interlocking connections (e.g., undercuts or perforations) mechanically support the composite strength [2]. Depending on the sprue position and interface geometry, it can be distinguished between butt and overlapping joint [12]. The joint design affects the molecular orientations in the interface (see Section 1.4.1.5). Molecules oriented parallel to the interface can participate in the diffusion process at higher temperatures, and contribute to enhancing adhesion. Diffusion processes can take place as a consequence of the first component melting due to the heat input of the second component. In butt joint connections the melt of the second component contacts the whole interface area approximatly at the same time. In overlapping connections, an overflowing occurs, which leads to different temperature and pressure conditions in the connection area. This can lead to local differences in adhesion. Furthermore an overflow leads to wall-sticking at the first component which affects the molecule orientation [12]. In most cases, the implementation is done stepwise, shafted, or butt, and can result in varying results for the according material combinations [6]. Therefore, a general statement cannot be made concerning the influence.

In addition to interlocking connections, a form fit is usually included in practice [2]. A form-fitted version is the most reliable option, even if it is more costly due to the need for a more complex mold. Moreover, the difference in modulus and the elasticity of the materials must also be taken into account. Differing deformation behavior in components can lead to a separation of the materials when load is applied.

1.4.3.3 Surface Roughness

There is the option of roughening the contact surface. One example of this would be molding the first component on a rough surface of the slider, which opens the volume for the second component [12]. This essentially enhances adhesion due to the larger contact surface, but is also connected to negative factors of influence. High levels of roughness can inhibit the wettability and enable air or dirt to enter the contact area [1,4,17]. A complicated penetration of the melt can lead to stress peaks in the interface [6]. Moreover, the larger composite surface can result in greater heat dissipation, which, in turn leads to a lower heat level in the contact area. However, in most cases, the influence of the roughness is much lower than the influence of process control and should be assessed depending on the material [1,9].

1.4.3.4 Mold Design

The mold design can significantly affect the processing conditions. Flow path, flow cross section, and the direction of the inflow define the local melt front speed, flow, and cooling conditions. In turn, these factors affect the compression ratio and temperature conditions, and thus also the viscosity. In addition, the sprue type and position play an important role. In the end, these factors influence the relaxation, orientation, and mobility of the molecular chains.

For this reason, it is difficult to make general statements about the influence of the mold design on the composite strength. Due to the large number of influencing factors, the processing of identical plastics in different molds can result in differing composite strengths [2].

In general, of course, the elimination of air is recommended in the contact area in order to counteract air pockets. Furthermore, thinner wall thicknesses in semicrystalline materials prove to be mostly beneficial with regard to the composite strength. This effect is supposed to be a consequence of the lower crystallinity of the first component induced by faster cooling [7].

The local pressure and temperature conditions vary in the component. Thus, the length of the flow path also has great significance for the composite strength. For a good adhesion, the temperature in the boundary area, the pressure, and the time of contact are recommended to be high. These conditions can normally be realized more easily near the gate, but they also strongly depend on the component geometry. Experiments have shown that the longer the flow path, the greater the composite strength. This is due to the so-called pressure delay time—meaning the time between wetting and building up pressure. This time decreases throughout the length of the flow path. At the end of the flow path, the pressure increase takes effect directly after wetting, which is beneficial for the composite strength. Near the gate a solidified surface layer can be developed, which is pressed on the first component with low pressure – at the beginning of the injection phase only the filling pressure. The pressure increase that occurs later, after the mold has been completely filled, does not have as much effect near the gate because the molecular mobility is already restricted in this area. But also, frontal injection attempts have shown that the composite strength is not determined by the pressure delay time alone [2,6,12,25].

1.4.3.5 Thermal Expansion Behavior

Varying degrees of thermal expansion in the components can lead to tensions in the interface that are mostly larger than flow-induced stresses. In hard-soft combinations in particular, expansion inhibition can occur in the surface area of the soft component because of the fixed connection to the hard component; thus, stresses can be induced [2].

Stresses negatively affect the composite adhesion. Similar heat contractions in both components are optimal, meaning, the closer the ratio of the thermal expansion coefficients of both applied materials is to 1, the better [7,15].

1.4.3.6 Injection and Holding Pressure

The processing pressures affect the wetting behavior, the molecular convergence, and the molecular orientation, and thus, the diffusion processes in particular.

The diffusion process itself is not dependent on the pressure [12]. Even though a higher composite strength is observable up to a certain pressure level, it can be traced to a reduction of cleavings between the joining partners [12]. The influence of the holding pressure is often regarded as subordinate in the literature [e.g., 1,2,8,12,13,25]. The effective direction of the holding pressure's influence is connected to the material combination and the injection sequence. It can be either positive or negative and is usually only measurable to a limited pressure value. Once this level of pressure has been reached or passed, macroscopically there is complete contact and a sufficient convergence for diffusion processes for both components [6,9].

The pressure-time-sequence of the holding pressure determines the shrinkage to a great extent. In this regard, a difference is observable concerning whether low pressure was applicated for a long time, or a high pressure applicated in a short period. Constant pressure loses influence over time due to continuing cooling. If the pressure is too high, molecular stretching, deformations in the contact surface, and local shrinkage and stiffness differences can result. They, in turn, negatively affect bonding strength [1,9].

1.4.3.7 Injection Speed

The injection speed essentially determines the degree of molecular orientation and, thus, affects the adhesion, orientation, and shrinkage [2].

With regard to the composite strength, also both positive and negative effects are possible [2]. A higher shear heating (higher mass temperature) and shorter pressure delay time that correlate with high injection speeds can increase composite strength [4,6,13]. At the same time, it is also possible for a higher molecular orientation to cause shrinkage-induced stress in the contact area, which, in turn, can inhibit bonding [2].

Due to the influence of the holding pressure, the effects of the injection speed are assessed in correlation with the material combination and further boundary conditions in differing ways in the literature [9]. While several investigations did not identify any significant influence [4], others observed a semiimportance of the composite strength at a greater injection speed [8,25]. However, it is crucial to note that in some material combinations, an influence was verified only in combination with high mass temperatures.

1.5 Machines and Processes

Having extensively elaborated the fundamental effects that play a role when manufacturing plastic-plastic composites using injection molding in the previous sections 1.2–1.4, this section now puts these effects into a practical context. Numerous, special injection molding techniques exist for the manufacturing of plastic-plastic composites. An attempt to structure these techniques often proves to be difficult. In most cases, a clear definition of processing terminology is missing, meaning the terminology used varies and can denote different processes depending upon the provider or sector. Moreover, it is also common that varying terminologies are used for nearly identical methods. This is often the case when several providers market similar methods with differing names. Due to overlapping with regard to content, establishing a hierarchical structure is complicated. Consequently, the existing approaches concerning this matter have resulted in varying outcomes [26–29].

This chapter has a different type of structure in an attempt to avoid this problem. Manufacturing methods for plastic-plastic composites can be distinguished by means of four criteria, which comprise the employed material combination, the point on which the components are combined, the chronological sequence, and the material and handling technologies utilized. Each special technique can be described as a combination of characteristics of these criteria. If you employ a morphological box, each special technique represents a path through the box. Figure 1.12 emphasizes this relationship and provides an example for the monosandwich method which is highlighted by the boxes with a dark background.

<table>
<tr><th>Criteria</th><th colspan="7">Realization</th></tr>
<tr><td>1. Material combination</td><td colspan="4">Compatible materials</td><td colspan="3">Incompatible materials</td></tr>
<tr><td>2. Point of material joining</td><td colspan="2">Dosage unit</td><td colspan="2">Barrel in front of the screw</td><td colspan="2">Runner system</td><td>Cavity</td></tr>
<tr><td>3. Injection sequence</td><td colspan="4">Sequential</td><td colspan="3">Simultaneous</td></tr>
<tr><td>4. Tool technologies</td><td>Standard tooling technology</td><td>Slider technology</td><td>Transfer technology</td><td>Rotary disk</td><td>Index plate</td><td>Linear tooling systems</td><td>Horizontal turning technologies</td></tr>
</table>

Figure 1.12 Morphological box to distinguish special molding techniques for plastic-plastic composites.

In the ensuing section, these four criteria mentioned in Figure 1.12 are introduced and their correlating parameters and effects are explained. In some cases, specific special methods are used as examples.

1.5.1 Material Combination—The Example of Assembly Injection Molding

The question of whether plastics of the same type are combined with one another, or whether two differing material types are combined, not only influences the later molded part properties (warpage, orientations in the interface, etc.), but also the process design. Materials are classified as the same type when they consist of the same basic material and do not contain any additives that can alter their properties strongly.

When combining identical materials, a good interface compatibility can generally be assumed because the parameters of significance for adhesion (i.e., solubility and polarity) are high, and the interfacial free energy equals zero (see Section 1.4.1).

By combining different materials, these conditions are not fulfilled automatically. The ratio of the surface stresses, as well as parameters such as the solubility and polarity, must be taken into account for each material pairing. Moreover, it is important to keep in mind that differing shrinkage behavior can promote the creation of interface stresses (see Section 1.4.3). This also can occur when employing the same materials with differing filling contents, or when there are greater temperature differences between the individual components (see Section 1.4.3).

In addition to a theoretical observation, practical results can also be employed. It is crucial to note that the created compatibility tables are not

	ABS	ASA	CA	EVA	PA6	PA6.6	PBT	PBTP	PC	PC/ABS	PC/PBT	PC/PBTP	PC-PET	PE-HD	PE-LD	PET	PETP	PMMA	POM	PP	PPO	PS	PS-HI	PSU	PVC-H	PVC-W	SAN
ABS	3																										
ASA	3	3																									
CA	3	3	3																								
EVA	3	3	0-1	3																							
PA6	2-3	2-3			3																						
PA6.6	2-3	2-3			2-3	3																					
PBT	3	3	3			3	3																				
PBTP	2-3	2-3	2-3		2-3	2		3																			
PC	2-3	2-3			2-3	1	2-3	2-3	3																		
PC/ABS	3	3			2-3	2-3		2-3	3	3																	
PC/PBT	2-3	2-3			2-3	2		2	2-3	2	3																
PC/PBTP	3	3			3	3		3	3			3															
PC-PET	2	2			2	2		2	2	2	2		3														
PE-HD	1	1	1	3	1	1-2	0	1-2	1-2	1	1	1		2-3													
PE-LD	1	1	1	3	1	1-2	0	1-2	1-2	1	1	1		3	3												
PET	3															1											
PETP	2-3	2-3						2-3	2-3	3	2	3	2	0-1	1		3										
PMMA	2-3	2-3						1	1	2-3	2-3	3	2	0-1	1		1	3									
POM	1-2	1	0-1		1	1		1	1	1	1	1	1	0-1	1	2	1-2		3								
PP	1	1	1	3	0-1	1	0	1-2	1-2	1-2	0-1	1	0	1-2	3			0-1	1	2-3							
PPO	1	1	1		0-1	0-1	0	1	1	0-1	1	1		1	1		1	0-1	1	1	3						
PS	1	1	1	3	1	1	0	1-2	1	1-2	0-1	1	0	1-2	1		1	1	1-2	1	3	3					
PS-HI	1	1-2	1-2	2-3	1-2	1-2	0	1-2	1-2	1	0-1	1	0	1-2	1-2		1	1-2	1-2	1-2	3	3	3				
PSU	3	3						3	3	2-3		3		1	1		3			1	1	1-2	1-2	3			
PVC-H	3	3																3		2		2	2		3		
PVC-W	2-3	3	3	1-2			3	3	3	3	3	3		1-2	1-2			3		1-2	1-2	1-2	1-2		3	3	
SAN	3	3	3	3	2-3	3	3	2-3	3	2-3	2-3	3	1	1	1-2	3		3	1-2	1-2	1	1	1-2		3	3	3

0 = no adhesion
1 = poor adhesion
2 = good adhesion
3 = very good adhesion

Figure 1.13 Adhesion compatibility of various material combinations for orientation purposes [2,8,16,26–28,30–33].

based on experiments that were carried out using standardized conditions. Therefore, they can be utilized only for reference purposes. Also, the high dependency of the composite adhesion on processing and geometrical parameters also puts the informative value of such compilations into perspective. Nevertheless, Figure 1.13 depicts a summary of several diagrams for orientation purposes.

However, substance-to-substance bonded connection is not always the goal of this process. Assembly injection molding aims for something that connects the components but it does not fixate them; thus, the components are still allowed to move against one another. Depending on the application, friction between the assembled parts is specifically desired, or else to be avoided. Typical reasons for utilizing assembly injection molding include the reduction of the number of assembly processes, the creation of permanent bonds, the manufacture of bonds that cannot be assembled using a conventional strategy, the achievement of closer tolerances, or the creation of abradant connections, such as friction or sliding clutches [34]. An incompatibility of the joining partners in accordance with the previously observed aspects is beneficial in this context. Nevertheless, combinations of identical plastics (meaning those with an essentially high interface compatibility) can be joined with one another by means of assembly injection molding if the shrinkage effects are specifically utilized.

In the case of such composites, short-term shrinkage plays a decisive role. Depending on the material or processing method, this can represent a large proportion of the processing shrinkage. On average, design engineers can expect short-term shrinkage of approximately 40%. However, this value can range between 15% and 60%. A reduction of short-term shrinkage can primarily be achieved by increasing the holding pressure, the tool temperature and the wall thickness. In addition, a longer holding pressure time, a shorter cooling time, or an increased mass temperature can make a minimal contribution to the reduction [35].

Regardless of the compatibility of the joining partners, melting and deformation of the components first injection molded should be avoided by selecting a moderate pressure and temperature level [36,37]. Often, a material with a higher melting temperature is used for the premolded part in order to guarantee that the processing window for the completion of the second injection process remains relatively wide [36].

1.5.2 Point of Material Joining

When manufacturing composites using injection molding, there are various methods to choose from for combining the materials. The most commonly used method is a joining of the components in the injection molding tool. Joining can also be completed in the nozzle, in the barrel in front of the screws, or when feeding in the material. In the next sections, these four methods will be illustrated with examples.

1.5.2.1 Joining Materials in the Tool—The Example of Bi-injection

For this method of joining, the availability of several injection units or the insertion of premolded parts is a prerequisite. In this case, primarily sequential processes are available (see Section 1.5.3). Bi-injection technology represents an exception. When employing bi-injection, two melts from two different injection units are simultaneously injected into the tool via two separate sprues. Figure 1.14 illustrates this.

The interface that develops during the joining of the materials, even it is fairly repeatable, cannot be clearly defined with regard to its contour. Merely a relocation is possible when adjusting the injection speeds of the single units. There are two crucial reasons that this method is selected. The first was that it is a simple process, with low tool costs. For this reason, it is used for applications with low optical requirements regarding the weld line. An example of such an application consists of illuminated multicolor icon switches in the automobile industry.

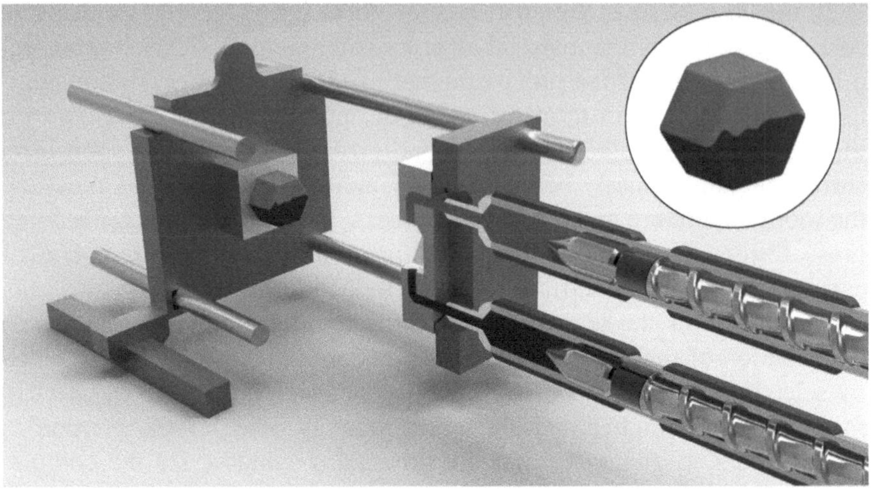

Figure 1.14 Bi-injection technology.

The second reason that this method is selected is because it enables a perfusion of the weld line by means of precise control of the injection and holding pressure processes. In particular, the special case of push-pull injection molding aims to do this. Here, a multiple perforation of the weld line is achieved by alternating injection [9,38].

Owing to the extremely high interface temperature in these processes, excellent composite adhesion can be achieved (see Section 1.4.2). Especially when using push-pull injection molding, it is possible to completely remove the orientations in the weld line. This factor, in combination with similar temperature levels of the melts, usually leads to low interface stress (see Section 1.4.2). In most cases, primarily identical materials are processed, which usually just differ in color. In this scenario, a high interface compatibility can be expected (see Section 1.4.3).

All other methods that join the materials in the tool consist of a sequential injection of the components. As a consequence, a lower interface temperature results. It is possible for melting to occur in the interface due to the material and processing parameters, however, it is not sufficient to significantly alter the orientations in the interface. Also, due to the larger temperature differences, the interface stresses are greater.

1.5.2.2 Joining the Materials in the Nozzle or in the Barrel in Front of the Screws—The Sandwich Technique

When materials are joined in the tool, there are several injection points. However, when the materials are combined in the nozzle, all materials are inserted into the tool via the same runner. As a consequence, the rheological parameters have a greater significance during the formation of the composite. The interface factors described in Sections 1.4.1–1.4.3 are still present, but, in relation, their relevance is less. Parameters such as the quality of the interface and the volume ratio of the individual components gain in significance with regard to the component.

The most important example of a combination of the melt in this manner is the sandwich method. Like the monosandwich method, it aims to create a multilayered structure. Both methods are classified as co-injection [27]. Figure 1.15 shows the differences found in the processing sequence.

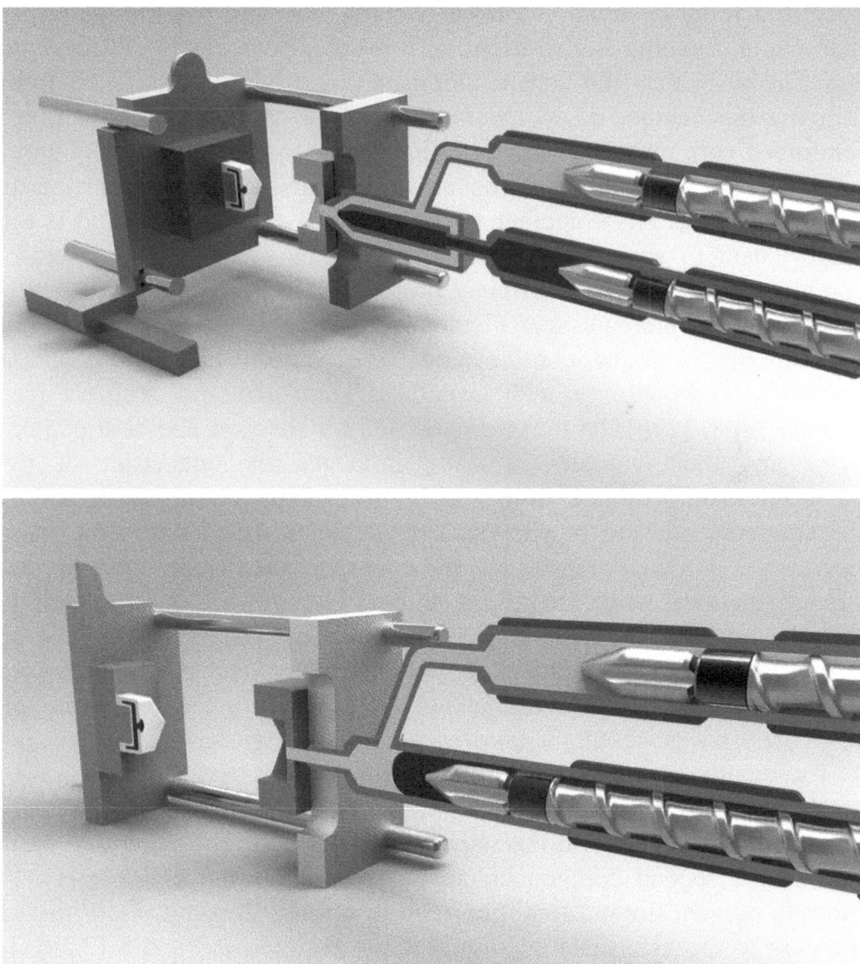

Figure 1.15 Traditional sandwich technique (top) and monosandwich technique (bottom).

Due to the similarity of both versions of the sandwich method, several aspects that are valid for both methods will be elaborated. Subsequently, a differential observation of the method variants will be carried out.

1.5.2.2.1 Sandwich Technique—General Information

The multilayered structure comprises a skin component situated on the component surface, and a core component surrounded by it. The utilization of such methods is beneficial for several reasons. A material with a low density or a foamed material can be used for the core component in order to achieve a lower weight. For wall thicknesses larger than 4 mm, foaming is also useful to prevent sink marks [2]. If recycled material is used as the core component, financial benefits result from utilizing this method [2,15]. Furthermore, the combination of a fiber-reinforced core material, with good mechanical properties, and an unreinforced skin material, with better surface and optical properties, is commonly used. Also, another potential of the sandwich method is the achievement of specific technical functions by means of purposefully combining materials with different properties [15,39].

The general prerequisite for composites is an interface compatibility between the skin and core material (see Section 1.4.1) [15], and a composite compatibility with regard to shrinkage and stiffness (see Section 1.4.3) [29]. The materials selected for the skin and core components and their volumes crucially influence the properties of the composites. For example, in the case of a combination of Polyamid 6 as the core material and polybutylene terephthalate filled with 20% glass fiber as the skin material with a larger core ratio, the flexural modulus and strength increase, while the impact strength goes down [40]. Accordingly, the adjustment of the ratio between the skin and core material must be devoted a high amount of attention. Decisive factors include the method of injection, the position of the sprue, and the design of the molded parts [2]. In contrast, the processing parameters are far less important [2].

The quality of the layered structure (more precisely, the precision in the layer's uniform thickness) is primarily determined by the viscosity of the components due to the sequence of processing. During the injection of the second component, a laminar flow is induced [39,41]. In order to prevent flow instabilities, the ratio of the shearing viscosities of the core to the skin material should range between 0.5 and 5 [2,29]. If the ratio is lower, ruptures occur; if it is higher, poorer filling behavior and a varying wall thickness distribution results [40].

A core material with a low viscosity changes direction slightly when faced with varying flow resistances. Thus, preferred flow paths are created, and, simultaneously, displacement no longer takes place in some areas. This effect, which is subject to self-enhancing, is referred to as the finger effect [26]. An observation of selective table values of viscosity is insufficient in predicting this effect, as the viscosity varies indirectly through time and local variables, or also through direct shear and temperature properties [29].

As was the case for the ratio between the skin and core, the mold geometry also plays an essential role concerning the quality of the layered structure. Rotationally symmetric mold parts are especially suitable for sandwich applications. The larger the divergence in the mold part design, the higher the probability that inhomogeneities will arise, and, in turn, the lower the ratio of the core material will be [2,39].

1.5.2.2.2 The Basic Sandwich Method

The regular sandwich method involves a joining of the skin and core material in the nozzle. Small injection weights can be integrated on standard machines by means of using a sandwich adapter plate, which is mounted on the side of the nozzle between the clamping platen and the tool [26]. In the case of larger components, or strongly differing injection weights for individual components, hot runner co-injection heads are employed [42]. The injection units are usually situated next to each other or on top of one another [26,42].

It is possible to employ both cold and hot runners. In the case of hot runners, a special manifold replaces the injection head [27]. Special nozzles are employed to control the melt flow of the skin and core components. They are characterized by separate valves for the individual components [26]. With regard to the mold itself, specific adjustments are not necessary [27].

Employing two separate injection units enables the precise control of the injection sequence. This sequence consists of three phases. In the first one, the skin material is injected, followed by the core material in the second phase. If sealing the gate is necessary, due to visual requirements, the third phase completes the sequence with another injection of the core material. Otherwise, the core material is used to compensate shrinkage in the third step. Usually, a simultaneous phase in which the skin and core material are injected at the same time is included between the steps to avoid abrupt changes in flow velocity and resulting flow marks [26,27,42].

1.5.2.2.3 Monosandwich Method

The monosandwich method specifically combines the materials in the barrel in front of the screw. An extruder that directs one of the components into an injection unit from the side is required. It is connected via a force-controlled, three-way valve [43]. As can be seen in Figure 1.13, the actual injection process is carried out by only one unit, meaning that switching between the units during the injection sequence is not necessary [43]. This, on the one hand, simplifies process control; but on the other hand, it simultaneously reduces the possibilities to influence the process. In order to prevent ruptures, the core content or injection speed may have to be reduced [40].

Here, the nozzle technique is less complex than in the regular sandwich method because it must be designed for only one melt flow [2,43]. The achievable ratio of core material is comparable to that found in regular sandwich injection molding, and it equals approximately 60% [39]. It is possible to use standard tools; however, it is not possible to use hot runners with a simple needle valve due to the mixing effect that occur [2,15,44]. Also, a sprue sealing cannot be accomplished by injection of the skin material when using the monosandwich method [39,45].

1.5.2.3 Combining Materials via the Material Feed—The Marbling Technique

In the previously discussed methods of joining, at least one of the components was in a molten state. Here, however, the materials are partially combined in a solid state, as can be seen in Figure 1.16. Even if the according methods are categorized into reproducible and nonreproducible variations [26], the mixing (and thus the appearance of the part as well) are subject to stochastic fluctuation.

Nonreproducible effects are primarily categorized under the term "marbling". Here, similar (but not identical) optical effects are aimed for in the components. These can be achieved by means of an inhomogeneous color distribution, or by combining different materials [26]. The materials should not be allowed to dissolve completely into one another (see Section 1.4.1.6). In particular, material combinations that possess the required interface compatibility for adhesion (see Section 1.4.1) but differ with regard to their melting behavior are highly suitable [26]. The lowest degrees of mixing are achieved by employing plunger injection units [26].

Reproducible variants display specific analogies to the monosandwich method. They can be achieved either by employing a complex

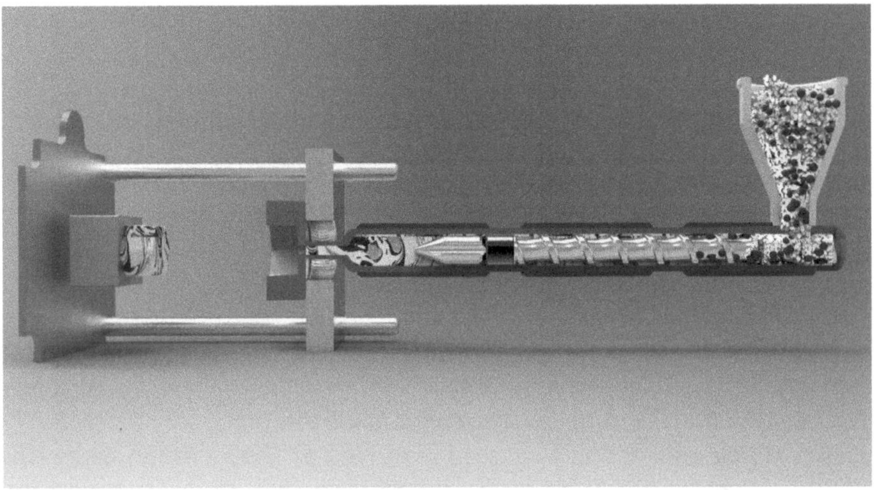

Figure 1.16 Marbling via the material feed.

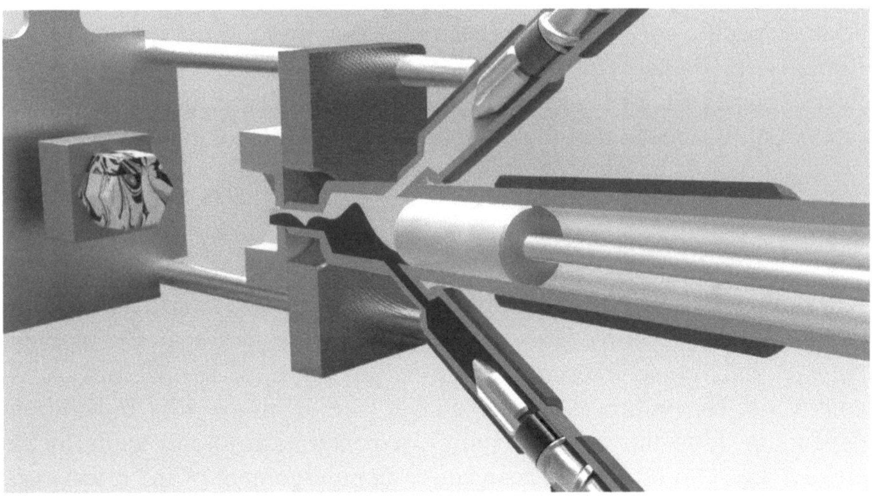

Figure 1.17 Marbling technology based on the loading of a plunger injection unit.

nozzle technique in a sequential injection, comparable to the regular sandwich method [26], or by the monosandwich method by loading a plunger injection unit in layers [26]. The latter technique is illustrated in Figure 1.17.

The close relation between the method's variables is clear when it comes to the Admix method. It is used to create sandwich structures and does not require additional plasticization units. Special dosing units alternating feed granulate of the core and skin material starting and ending with the skin component. As a result a layered structure of melts is generated in the barrel in front of the screw. However, achievable core-skin ratios are lower than in the case of the classic sandwich technique [26]. All marbling technologies can be carried out on conventional cold runner tools and do not require special tool modifications [46].

1.5.3 Injection Sequence

The chronological order of the injection processes strongly influences the later properties of the components. First and foremost, a differentiation can be made between simultaneous injection and sequential injection sequences. The time between each injection process is the decisive factor. When injecting in a successive circuit, as shown in the sandwich-molding process, the time between each sequence equals almost zero. Breaks that last several seconds (common in many rotational tool concepts), or up to several days (insertion technique) are possible. In several specific cases (i.e., sandwich injection molding), a mixed form is formally available. This is because both simultaneous and sequential injection are parts of the cycle. In the next section, the effects of various orders on the adhesion in the composites will be elaborated.

1.5.3.1 Simultaneous Process Control

Simultaneous process control implies that the viscosities and the temperatures of the components are on comparable levels at the moment the materials come into contact. Often, an interface temperature sufficient for diffusion processes is available (see Section 1.4.2). In addition, depending upon the process control, a penetration of the melts in the contact area can occur. This can cause an enlargement of the effect contact area, but also a reduction of the surface stress differences of the flow front (see Section 1.4.3). The orientations in the material that are otherwise mainly parallel to the interface, especially in fiber-filled materials, do not develop as well when melts collide. Thus, this phenomenon, also allocated to surface compatibility (see Section 1.4.1), can positively influence composite adhesion.

If the difference in temperature between the components is low, the shrinkage-based interface stresses are merely based upon deviations in the heat expansion of the materials (see Section 1.4.3). Therefore, they are lower than in sequential process control.

1.5.3.2 Sequential Process Control

If one component is injected and then another injected after that, then it can be assumed that the available interface temperatures are less than those found in simultaneous process control. However, this does not imply that diffusion processes across the interface are impossible. The prerequisite here is achieving the necessary temperature limit on the surface of the component injected first (see Section 1.4.2). A very short period between injection of the first component and contact with the second is beneficial (see Section 1.4.2). Moreover, differences in the processing temperatures of the materials can be specifically utilized. For this purpose, the material with the lower melting temperature should be used as the first component. The manufacture of hard-soft components is an exception because deformations of thermoplastic elastomer (TPE), which usually melts at a lower temperature, inhibit its use as a first component. If it is not possible to achieve interface temperatures higher than the temperature limit for diffusion processes of the first component due to processing (i.e., in insertion processes), adhesion processes gain significance (see Section 1.3.1).

1.5.4 Tool Technologies

Different injection molding tool concepts are available in practice which can be used to manufacture plastic-plastic composites. However, no clearly defined, standardized classification exists. Various attempts to classify [e.g., 4,6,8,46,47] utilize similar criteria to differentiate between the tool technologies. Therefore, the technologies are distinguished, based on whether the preformed piece, which was made in the first injection process, remains in its current position in the tool or is repositioned. In the case of the latter, a differentiation can also be made between the types of repositioning. Repositioning can be achieved by either using handling machines or moving the tool, which can have either a rotational or linear design. Figure 1.18 shows an outline of the classifications mentioned here. Implemented tool technologies have been used as examples.

1.5.4.1 Slider Technology

This tool technology is often referred to as the core back technique in accordance with the method associated with it. The process starts with the injection of the first component. After a short cooling period,

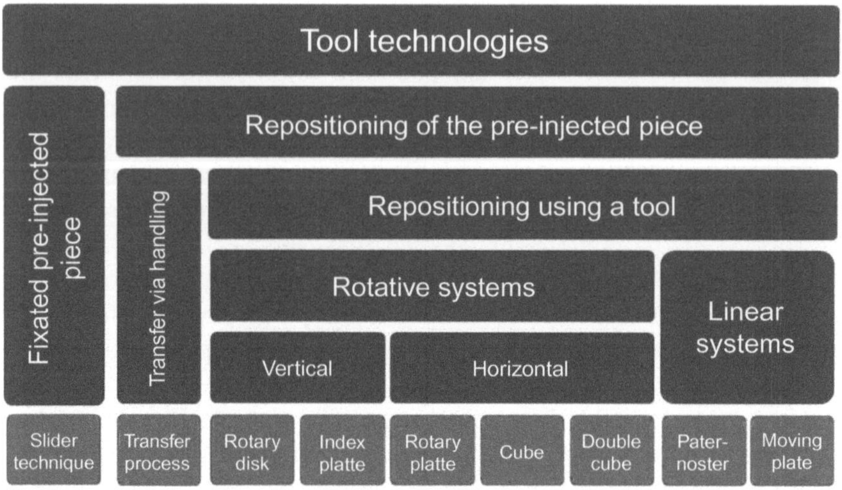

Figure 1.18 Classification of tool technologies.

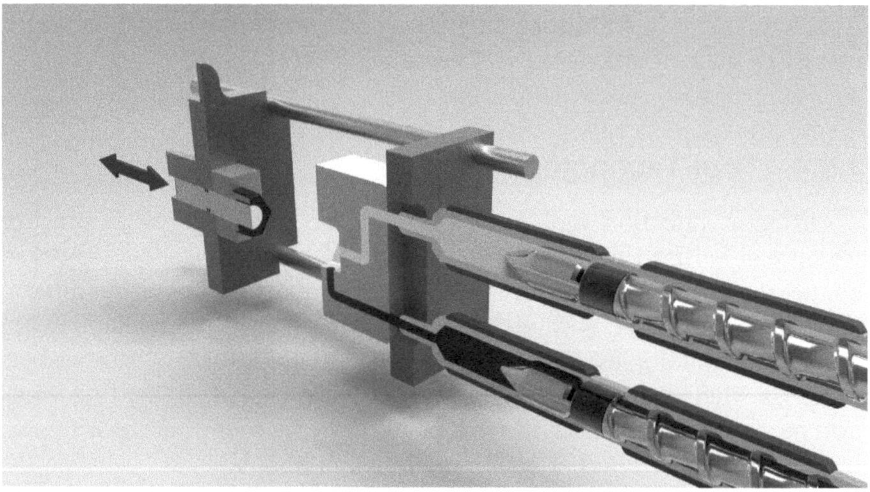

Figure 1.19 Core back technique.

the cavity volume is enhanced for the second component by means of a core or slider. Figure 1.19 depicts this principle.

Tools based on this concept often have a very simple design [46]. Consequently, their structure is compact, which, in turn, enables an effective utilization of the available construction space [42]. Moreover, the investment costs are low in comparison to that of the methods that will be introduced later in this chapter [7,30,42]. The simple design also has disadvantages though. The point of injection can be chosen flexibly

[46], but the cavity design on the ejection side is highly restricted because additional space must be provided by means of an axial movement. In certain cases of application, a slider must be applied on the nozzle side [48]. Complex geometries can be achieved using collapsible cores; however, this is usually not the best method from an economic perspective [6].

The cycle times that can be accomplished with slider tools are longer than for rotating tools. Even though a cavity opening between the injection of the first and second components can be discarded [26,46], the sequential process requires more time for the injection of successively injected units [6,7,46,48] than simultaneous injection processes in other methods. Due to the low tool cost, but higher cycle time, this method is primarily used for small series [48]. With regard to composite adhesion, slider tools prove to have advantages that repositioning technologies do not. Owing to the quick succession of the injection steps, the preinjected piece does not cool off as much, and higher interface temperatures are achieved (see Section 1.4.2) [6,46]. Also, with regard to interface stresses, better conditions are to be expected.

1.5.4.2 Transfer Process

The transfer process consists of moving the preinjected piece into a second cavity by means of a handling machine. Two types of transfer are distinguished. In the case of relocating, the preinjected piece is moved from one cavity to the next within the tool [48]. The cavities can be either above one another or next to each other [46]. In the case of the insertion technique, the transfer is carried out between two injection molding machines [48]. The difference between both variations is illustrated in Figure 1.20.

Analogous to the slider technology, tools that employ transfer technologies in general require a simpler tool design than that of rotating or linear alternatives. Moreover, the compact structure enables good utilization of the available clamping surface [48]. The insertion technique, the transfer between two machines, also enables an efficient utilization of the clamping force [46]. The possibility to position the cavity in the center of the tool [4], as well as the potential to adapt the size of each injection molding machine to the injection process realized on it, have positive effects. Repositioning between machines can avoid the problem of thermal separation found in the combination of thermoplastic and elastomeric materials, because the temperature of the cavities can be controlled entirely separate from one another [6].

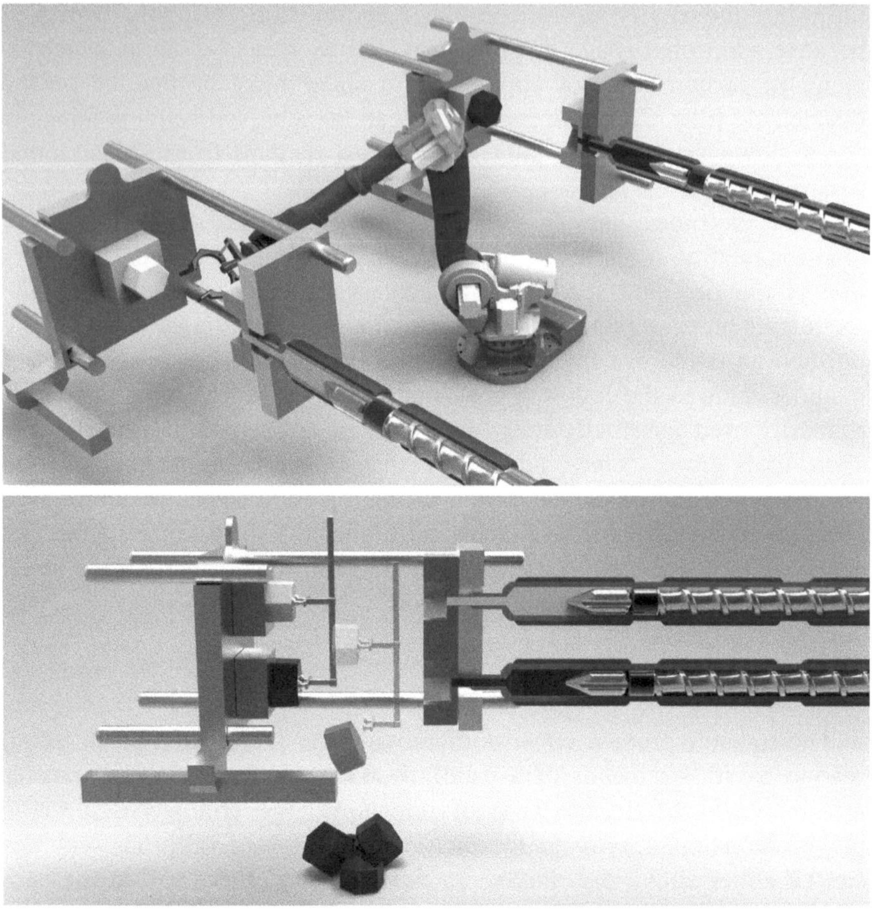

Figure 1.20 Relocating technique (bottom) and insert technique (top).

In general, flexibility in the repositioning process enables the greatest degree of design freedom of all the tool technologies discussed in this chapter [2,42]. This, however, has to do with the selection of the injection position; but there are limitations when it comes to creating delicate geometries, because handling can be difficult [46]. For this reason, primarily simpler and more robust contours are made [2,27].

Generally, positioning the components is connected to challenges during the handling process [46]. Short-term shrinkage must be reckoned with during the repositioning [2,7,27]. In particular, placing the component on cores is regarded as highly critical [7]. Semicrystalline materials do display a greater degree of shrinkage and are at higher risk for this problem. Amorphous thermoplasts are at greater risk of breaking during this process due to their brittle nature [7]. Besides problems during the actual insertion process, quality defects, such as warpage, can also occur later (see Section 1.4.3.1) [2,6].

Regarding the cycle time, simultaneous injection is possible[1] [46], yet repositioning with a handling machine usually requires more time than a technique such as repositioning using a rotating tool concept. The repositioning method is often used for larger quantities, while the insertion technique is usually reserved for smaller series and prototypes [6]. In addition to economic aspects, there are technical necessities that make the utilization of transfer methods imperative. For instance, this is the case for very large components and accordingly large tools, which cannot be moved on a rotary disk due to their mass (see Section 1.5.4.3) [6,7,27]. Even when components cannot be held on an index plate (see Section 1.5.4.4), the repositioning method is used [7]. The insertion technique is especially suitable for thermoplast-elastomer composites [2,42]. In cases where there are large geometry changes between the first and second component concerning the required tool size and clamp force this technology is favorable [30].

The longer repositioning times of the transfer methods negatively influence the interface temperature (see Section 1.3.2), and, thus, the adhesion as well. The larger temperature differences in combination with hardly reproducible cooling conditions can lead to increased interface stresses (see Section 1.4.3). One advantage of this technique is the ability to positively influence the cooling times and other processing parameters independently. This option is partially available for the repositioning method, and fully for the insertion method.

1.5.4.3 Rotary Disk

In tools with rotary disks, one side of the tool is rotated to move the formed component to the next station. The rotary plate is almost always located on the side of ejection, because if it were positioned near the nozzle, a rotating sprue would be needed as well [6]. Depending on the number of components, the angle of rotation varies. Thus, for two components, the angle is 180°; for three, it is 120°, etc. In order to reduce the cycle time, one station of the rotary disk can be reserved for the demolding process in the closed tool. For this purpose, an opening is needed in the nozzle part of the tool [48]. Rotary disks can be a fixed

---

[1]In contrast to the previous section, *simultaneous* refers to the simultaneous action of several injection units at different tool positions, not the simultaneous filling of cavities.

Figure 1.21 Rotary disk.

part of the injection molding machine, but they also can be integrated into the tool [6]. The machine-integrated version is more common in larger machines, the tool-integrated version in smaller machines [42]. The power unit of the rotary disk is achieved either hydraulically or with servo motors [46] (Figure 1.21).

If the rotary disk is an integral part of the machine, the tool itself will have a simpler construction [7]. Moreover, the rotary disk can be used for various tools [7,48]. Because the molded parts are securely anchored in the rotating cavity, precision when positioning is not a challenge compared to transfer or index tools [6].

One disadvantage of the rotary disk technology is its need for a large amount of space. Due to a higher mold height, tie bar extensions of 200 mm and more are needed [46]. The tool width is also larger and requires more space between the tie bars [49]. Values ranging between approximately 50–100 mm are common [46]. The ejection plates turn with the tool and therefore must be present at every station [48]. With regard to the component geometry, there are limitations because an enlargement of the cavity volume in the second cycle or later cycles can be done only via the mold contour on the nozzle side [6,27].

Most times, the projected area by the part geometry differs at the various stations of the rotary disk. The consequence is an asymmetrical distribution of the inflation [4,6]. In the case of larger differences, a design that includes 70%–80% of the maximum clamping force must be selected [47]. Moreover, the cycle and the limited thermal separation prevent an independent optimization of the single processes, meaning that compromises must be made with regard to the cooling and holding pressure times [4,6].

Challenges arise when dealing with larger molded parts. Due to the occurring tilting torque, the ability to utilize rotary tables on machines with a clamp force of up to 2000 tons is limited [49]. Also, the rotational movement itself poses quite the challenge in itself. The rotary feedthrough of supply lines, as well as the fact that size is limited due to stability issues [49], limit the dimensions of the hydraulic and cooling supply [46]. In addition, there are limitations concerning the thermal separation at the rotational feedthrough segment [49].

Thanks to their high rotation speeds and simultaneous injection at the stations, rotary disk tools enable short cycle times to be achieved [26]. When this advantage is combined with a design freedom that is larger by far than for the slider technology [6,27] the rotary disk becomes one of the most commonly used multicomponent tool technologies [30].

The achievable interface temperatures depend upon the theoretically required cooling times of the components injected at the individual stations. Should these diverge strongly from one another, then some components will cool down excessively, and the interface temperature (see Section 1.4.2) during the following injection step will be lower. In addition, an increase in the interface stress should be expected (see Section 1.4.3). When cooling times are similar to one another, the fast and reproducible rotational movements are beneficial.

1.5.4.4 Index Plate

The index plate concept is similar to that of the rotary disk. However, here, the entire ejection side of the tool is not rotated, but simply a plate, or a beam or cross. As was also the case for the rotary disk, the number of stations is not set. In addition, this concept enables stations to be used merely for cooling or demolding [46]. During a rotation process, the machine ejector and the attached index plate are extended [30], and then subsequently turned by one station and drawn in again. The rotating cores can be pulled behind the cavity, turned, and, subsequently, reinserted into the cavity from behind [6]. The rotational part of the movement is powered by a servo, or hydraulic motor [46]. Also, rotation axes are also possible outside the tool [6] (Figure 1.22).

Two designs are distinguished here: index plate tools, which have a completely rotating tool plate; and hub tools, in which only a beam or cross completes the rotational movement. The latter is sometimes also referred to as an index plate [4]. The components are kept on collapsible

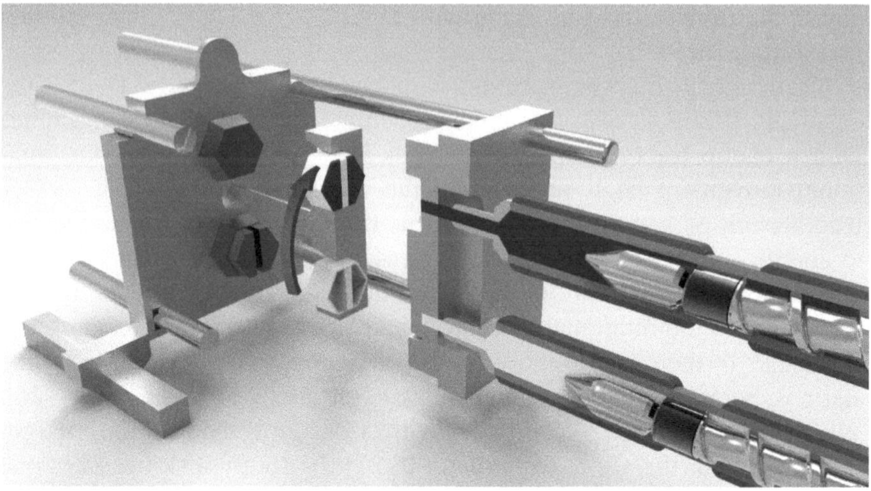

Figure 1.22 Index plate technology.

cores with undercuts [46,48]. Due to the low total mass, higher speeds can be achieved with lifting than with the rotating tool [46]. As the ejector plate remains unmoved, only one ejector system is required [48].

The problem of material insertion is also not prevalent here. In addition, geometries can be created to which a component will be added, either on the side of ejection [48] or on both the side of the nozzle and the side of ejection [7,27,30]. Depending upon the application, in particular when using rotating tools, hot runner systems are applicable in only a limited sense [46,48].

Utilizing index plate tools is suitable when only one product or a few products are to be manufactured on a machine. Thus, a machine-integrated rotary disk would not amortizc itsclf. Moreover, index plate tools are often utilized due to their large degree of component design freedom [7]. With regard to the attainable composite adhesion, comparable criteria apply as for the rotary disk tool.

1.5.4.5 Linear Systems

The simplest, linear system is the sliding table, which is a cavity plate located on the ejection side that moves between two or several positions vertically. At each position, there is an injection unit for the according component. Injection is sequential, meaning that only one injection unit is active at a time. Similar to the rotary disk tool, the standard method only allows a nozzle-sided cavity variation. However, in

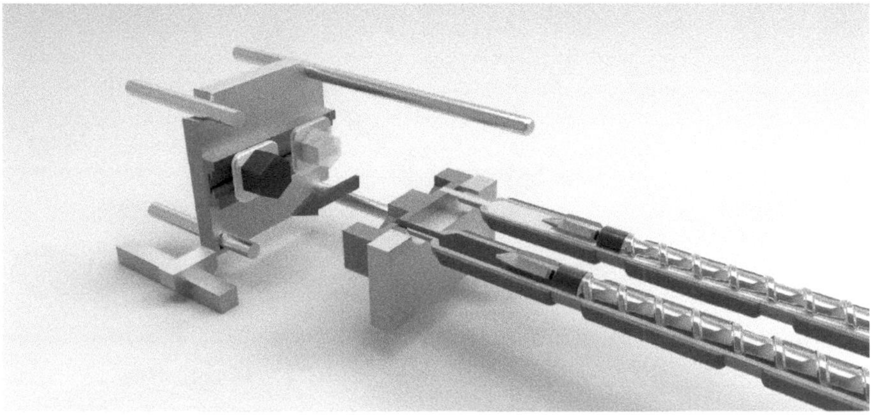

Figure 1.23 Sliding table.

the style of hub systems, there is a variation in which a central element loaded with cores is used, and which enables an adjustment of the cavity on the side of ejection. Thanks to its simple technology, a sliding table involves relatively low investment costs. The longer cycle times that result from the sequential processing are the reason that this tool concept is mostly employed for smaller quantities in smaller series [42]. The sliding table concept is depicted in Figure 1.23.

The Paternoster technology represents a concept that is eligible to compete with regard to the cycle time. Analogous to the elevator concept with the same name, the cavities on the same side of the tool are shifted sideways and continue downward in a linear motion until they finally rise again in the opposite direction. By means of employing several cavities, a simultaneous process cycle, as well as a demolding, are enabled in the closed tool. The movements of the cavities are screw-driven. The main benefit in comparison to rotary disk tools is the need for less space, meaning holm adjustments are not needed [46]. With regard to the composite adhesion, for both tool concepts, similar effects can be expected as those for rotary disk tools.

1.5.4.6 Horizontal Turning Technologies (Stack Turning Technologies)

While the previously described rotating technologies were based on a rotation on a horizontal axis, the technologies described in the next sections consist of tool rotations that take place on vertical axes. The concept utilizes a central plate that completes the turning movements.

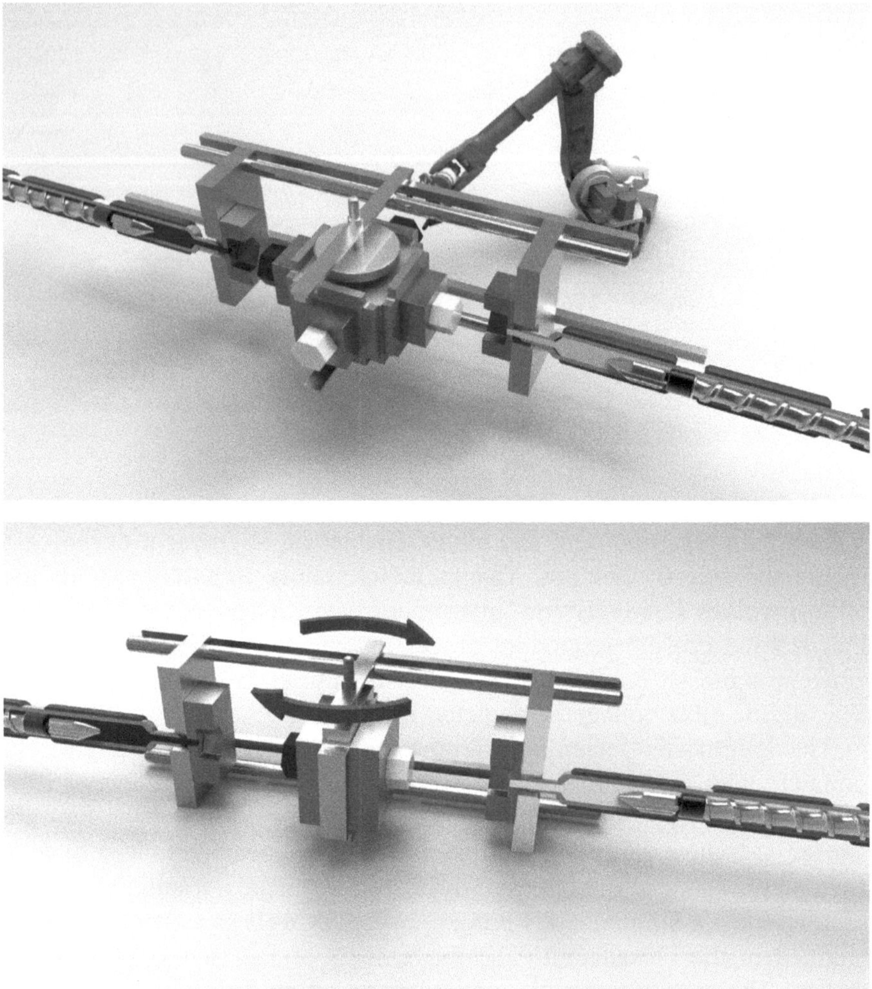

Figure 1.24 Cube technology (top) and stack turning plate (bottom) by comparison.

Depending upon the number of planned stations, there will either be a plate (two stations), or a cube (four stations). The variation with two stations is referred to as stack turning technology [4], while the variation with four stations is commonly called a rotary central block [27] or cube technology. These concepts are shown in Figure 1.24.

Perhaps the greatest advantage of this technology in comparison to others is that the clamping force provided by the machine can be utilized for both parting planes owing to their being arranged in opposite directions. Therefore, when employing two injection processes with identical clamping force requirements, the needed clamping force can be divided in half; thus, the machine size can be reduced [2,4,46,48]. The required amount of clamping surface reduces because, in contrast to rotary disk tools, the cavities of all stations do not all have to be positioned on one parting plane, but can be distributed over a second parting plane, which is located across too [4]. In cube tools, the additionally available stations on the sides can be used for intermediate steps, such as coating, cooling, demolding, or inserting of additional components [2,46,48]. The separate parting planes also enable injection compression processes, which are often employed for optical components [42,49]. Limitations arise, however, concerning the positioning of the injection units because the second injection unit must always face in the opposite direction of the first in the movable tool plate [48]. Although complex redirections of the melt can be avoided [49], a need for a larger space for the machine results.

Because the supply lines can be directly hooked up and lack a rotary feedthrough, faster processes with higher levels of required cooling can be realized using stack mold technologies [49]. From an economic perspective, cube technologies only prove useful when at least three stations of the cube are utilized, because their tool costs are comparably high [49]. The realizable geometries are comparable to those of the rotary disk tool, because a variation of the cavity is also only possible via one side of the tool (i.e., of the cube). Due to the advantage of the reduced clamping force and surface, the cube technology is predestined for flat and large components, or applications that require a high number of cavities [42,46].

Since at least one pause in the cycle between injecting the first and second components takes place in the four-station version, lower interface temperatures result (see Section 1.4.2), and, due to the thermal properties of the materials, mostly higher interfacial stresses occur (see Section 1.4.3). These have according effects on the composite adhesion. However, this intermediate step can be used to apply adhesion enhancing methods, such as plasma treatments. This, in turn, positively affects the composite adhesion by inducing increased interface compatibility (see Section 1.4.1).

In addition to the standard technology, further variations of the stack turning technology exist. Core segment rotation technology utilizes a center plate that is vertically divided into different segments, which each rotate on their own axes (Figure 1.25). One advantage is that smaller masses have to be moved when turning, and, owing to the smaller rotation radii, lower opening strokes are needed. Stations with either two stations or four stations are used [50].

In the case of applications where the assembly of the created components is to be carried out, there is the option of what is called double

Figure 1.25 Core segment rotation technique.

cube technology. Here, instead of merely one cube, there are two consecutive cubes situated between the tool plates. A molded part is created in each cube. The intermediate positions on the sides can be used to insert additional components. The final assembly step is carried out in the parting plane between the cubes. A large advantage of this method is that, in addition to process integration, that components from the same cavity pairs are always combined with one another owing to the higher number of cavities in the tools. With the correct settings, the reject rate can be considerably reduced [46,51].

References

[1] Brinkmann S. Verbesserte Vorhersage der Verbundfestigkeit von 2-Komponenten-Spritzgiessbauteilen. Dissertation, Aachen; 1996.

[2] Ehrenstein GW. Handbuch Kunststoff-Verbindungstechnik. Hanser; 2004.

[3] Michaeli W, Lettowsky C. Sonderverfahren des Spritzgießens. In: Medizintechnik-Life Science Engineering. Springer, Berlin, Heidelberg; 2009, p. 597–639.

[4] Kleeschulte R. Modellierung der Verbundfestigkeit von Hart-Weich-Kombinationen polymerer Werkstoffe, Dissertation, Paderborn; 2012.

[5] Bauer R. Der nächste Schritt in den Motorblock. Kunststoffe 2012;05:49–52.

[6] Kuhmann K. Prozeß- und Materialeinflüsse beim Mehrkomponentenspritzgießen. Dissertation, Erlangen-Nürnberg; 1999.

[7] Kaiser E. "Scale-Up" Methoden zur Prozessführung von 2K Spritzgussteilen im Hinblick auf ihre Verbundhaftung. Dissertation, Paderborn; 2012.

[8] Schuck M. Kompatibilitätsprinzipien beim Montagespritzgießen. Dissertation, Erlangen-Nürnberg; 2009.
[9] Kühnert I. Grenzflächen beim Mehrkunststoffspritzgießen. Dissertation, Chemnitz; 2005.
[10] Potente H. Fügen von Kunststoffen: Grundlagen, Verfahren, Anwendung Hanser; 2004.
[11] Fourche G. An overview of the basic aspects of polymer adhesion. Part II: application to surface treatments. Poly Eng Sci 1995;35(12):968–75.
[12] Jaroschek CS. Spritzgiessen von Formteilen aus mehreren Komponenten. Dissertation, Aachen; 1994.
[13] Candal MV, Gordillo A, Santana OO, Sánchez JJ. Study of the adhesion strength on overmoulded plastic materials using the essential work of interfacial fracture (EWIF) concept. J Mater Sci 2008;43(15):5052–60.
[14] Lee L-H. Fundamentals of adhesion. New York, NY: Plenum Press; 1991.
[15] Potente H, Koch M, Moritzer E, Kaiser E, Ridder H, Schäfers M. Mehrkomponenten- Verfahren richtig anwenden, 25 Jahre KTP. Kunststoffe 2005;(2):44–7.
[16] Wunderle J, Michaeli W. Mehrkomponentenspritzgießen von Hart/Weich-Komninationen, Haftungsmechanismen und Haftungsprüfung. VDI Wissensforum Verbundspritzgießtechnik, Dietenhofen; 2012.
[17] Fourche G. An overview of the basic aspects of polymer adhesion. Part I: fundamentals. Poly Eng Sci 1995;35(12):957–67.
[18] Amesöder S, Kopczynska A, Ehrenstein GW. Multi-component processing, plasma makes for a strong bond. Kunststoffe International; 2003/09:31–34.
[19] Gleich H. Zusammenhang zwischen Oberflächenenergie und Adhäsionsvermögen von Polymerwerkstoffen am Beispiel von PP und PBT und deren Beeinflussung durch die Niederdruck-Plasmatechnologie. Dissertation, Duisburg; 2004.
[20] Kopczynska A. Oberflächenspannungsphänomene bei Kunststoffen—Bestimmung und Anwendung. Dissertation, Erlangen-Nürnberg; 2008.
[21] Brewis DM, Mathieson I. Flame Treatment of Polymers to Improve Adhesion. In: Adhesion Promotion Techniques. Technological Applications. Materials engineering, New York; 1999, p. 175–190.
[22] Buchmann A, Dodiuk-Kenig H. Laser surface treatment to improve adhesion. In: Adhesion Promotion Techniques. Technological Applications. Materials engineering, New York; 1999, p. 205–244.
[23] Uehara T, Corona discharge treatment of polymers. In: Adhesion Promotion Techniques. Technological Applications. Materials engineering, New York; 1999, p. 191–204.
[24] Wertheimer MR, Martinu L, Klemberg-Sapieha JE. Plasma treatment of polymers to improve adhesion. In: Adhesion Promotion Techniques. Technological Applications. Materials engineering, New York; 1999, p. 139–174.

[25] Kuhmann K. Verbundfestigkeit von Thermoplastkombinationen. Kunststoffe 2001;91(3):115–18.

[26] Johannaber F, Michaeli W. Handbuch Spritzgiessen. Hanser; 2004.

[27] Kamal MR, Isayev AI, Liu S-J. Injection molding, technology and fundamentals. Hanser; 2009.

[28] Stitz S, Keller W. Spritzgießtechnik, Verarbeitung—Maschine—Peripherie. Hanser; 2004: p. 2.

[29] Zipp T. Fliessverhalten beim 2-Komponenten-Spritzgiessen. Dissertation, Aachen; 1992.

[30] Eckhardt H. Mehrkomponenten- Spritzgießtechnik. VDI Wissensforum Verbundspritzgießtechnik Dietenhofen; 2012.

[31] Ehrenstein GW, Kuhmann K. Verbundfestigkeit beim Mehrkomponentenspritzgießen von flächigen Hart-Weich- und Hart-Hart-Verbindungen. In: Ehrenstein, G. W. und Kuhmann, K. Mehrkomponent-enspritzgießen. Technologie, Prozeß und Verbundeigenschaften. Ingenieur-Werkstoffe, VDI, Düsseldorf; 1997.

[32] Kneißl B. Hart-Weich-Verbindungen mit Thermoplastischen Elastomeren (TPE). VDI Wissensforum Verbundspritzgießtechnik Dietenhofen; 2012.

[33] Mitzler J. Neue Verfahren schaffen neue Möglichkeiten, Mehrkomponenten der nächsten Generation. VDI Wissensforum Verbundspritzgießtechnik Dietenhofen; 2012.

[34] Feulner R, Amesöder S, Kornsteiner W, Künkel R. Der richtige Dreh (Teil2), Reibschlüssige Verbindungen durch Montagespritzgießen als Überlastschutz.. J Plastics Technol 2009;5:5.

[35] Karlinger P, Bürkle E. Kurzzeitschwindung beim Mehrkomponentenspritzgießen. Kunststoffe 2001;91:54–9.

[36] Amesöder S, Feulner R, Drummer D, Künkel R. Der richtige Dreh (Teil1), Kennwerte für das Montagespritzgießen. Zeitschrift Kunststofftechnik 2008;4:4.

[37] Schmachtenberg E, Schuck M, Kühnert I. Urformen und Montieren in einem Prozess. Kunststoffe—Werkstoffe, Verarbeitung, Anwendung 2007;4:24–31.

[38] Sumitomo (SHI) Demag Plastics Machinery GmbH: Bi Injection. http://www.sumitomo-shi-demag.eu/de/verfahren/mehrkomponententechnik/bi-injektion.html.

[39] Jaroschek CS. In: Ehrenstein GW, Kuhmann K, editors. Verfahren und Anwendung des Monosandwichspritzgießens. VDI Düsseldorf; 1997.

[40] Seldén R. Co-injection molding: effect of processing on material distribution and mechanical properties of a sandwich molded plate. 2000;40(5): 1165–76.

[41] Jaroschek C, Steger R. Verbundspritzgießen mit einer Spritzeinheit, Das erweiterte Monosandwich-Verfahren. Kunststoffe 1999;89(7):38–9.

[42] N.N.: Mit Multinject-Technologie immer gut beraten Engineering, Mehrwert durch mehr Komponenten. Technical Information, Krauss Maffei; 2013.

[43] Das Mono-Sandwich-Verfahren: Neuer Weg zur Verarbeitung von Regeneraten. Plastverarbeiter 1993;44(7):14–17.
[44] EWIKON Heißkanalsysteme GmbH: Koaxialnadelverschlusstechnik. http://www.ewikon.com/de/produkte/spezialsysteme/koaxialnadelverschlusstechnik.html.
[45] Moritzer E. Beschreibung und Vorhersage der Kern-/Hautverteilung bei Sandwichbauteilen mit Hilfe der Ähnlichkeitstheorie, Schlussbericht der Forschungsstelle(n) Nr. 1, Verein zur Förderung der Kunststofftechnologie e.V. 2011.
[46] Pruner H, Nesch W. Understanding injection molds. Hanser; 2013.
[47] Rahnhöfer K. Prozess- und Werkzeugvarianten beim Verbundspritzgießen. Dietenhofen. VDI Wissensforum Verbundspritzgießtechnik; 2012.
[48] Rief B. Mehr Farbe und Funktion. Kunststoffe 2003;6:20–6.
[49] Handschke A. Vom Drehen zum Wenden, Wendeplattentechnik. Kunststoffe 2009;(10):78–82.
[50] Steinbichler G. Werkstoff und Verfahrenskombinationen im Spritzguss bestimmen die Zukunft. VDI Wissensforum Verbundspritzgießtechnik, Dietenhofen; 2012.
[51] Thümen T. Die vier Seiten eines Würfels. Kunststoffe 2012;(09):26–30.
[52] Michaeli W, Lettowsky C. Dünnwandig, leicht und verzugsarm. Mehrschichtige Spritzgussbauteile. Plastverarbeiter 2005;56(4):76–7.

2 Foam Injection Molding

Martin Rohleder and Fabian Jakob
Universität Kassel, Institut für Werkstofftechnik, Kunststofftechnik, Kassel, Germany

2.1 Why Foam Injection Molding?

Thermoplastic foam injection molding (FIM) is a highly complex process in which the interplay between process parameters and material properties largely influences the final component. This chapter focuses on the essentials for the creation of polymer foams and of the foaming process, and introduces various methods and tool technologies of FIM. The morphology and the mechanical and optical properties of the foams will be illustrated, and the correlation between the individual process parameters and the later foam structure will be elaborated.

2.1.1 General Idea

Thermoplastic foam injection molding is a special injection molding method that was originally developed to solve several problems that arise when injection molding compact components [1]. Various processors of plastics attempted to counteract the occurrence of sink mark defects in components by incorporating stiffening ribs created by adding small amounts of baking soda to the polymer melt. So, the first thermoplastic foams were developed from this initial idea [2]. Afterwards it was discovered that, in addition to reducing sink marks, foaming thermoplastics positively influences other quality characteristics of injection molded pieces, such as the warpage and internal stresses [2].

The wall thickness of components produced using a conventional injection molding process is generally limited to 1–5 mm. In terms of process technology, these thin cross sections require a high injection speed and high injection pressure to avoid premature freezing of the melt in the cavity [1]. During foam injection molding, the viscosity is somewhat lower due to the gases released in the polymer melt. This can result in longer flow paths in the cavity (Section 2.1.2). In contrast to injection molding compact components, foamed components do not require a holding pressure because the expanding gas functions as an

Specialized Injection Molding Techniques.
DOI: http://dx.doi.org/10.1016/B978-0-323-34100-4.00002-X

Figure 2.1 Monitor enclosure often manufactured using the thermoplastic foam injection molding method.

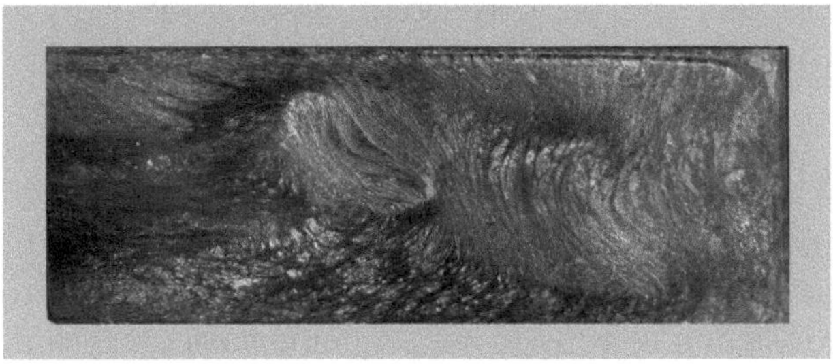

Figure 2.2 Formation of streaks on the component surface of thermoplastic foam injection molded components.

internal holding pressure in the produced component. Moreover, the gas pressure is preserved after the sprue has frozen, as well as throughout the entire cooling time. This, in turn, positively affects the dimensional stability of the components [2] (Figure 2.1).

Due to processing, components produced using foam injection molding display poor surface quality, and they often only obtain the desired quality after additional postprocessing [2]. The formation of streaks on the surface, as well as a relatively high roughness depth of 20–100 μm, are typical. These effects can be attributed to the tearing and collapsing of individual foam pores on the tool surface during injection [3] (Figure 2.2). They freeze on the tool wall, creating a well known swirled pattern. When the polymer melt cools down quickly on the tool wall, an unfoamed layer forms on the component surface because the low viscosity keeps the blowing agent from expanding. Thus, an integral foam structure is created, which has a compact skin layer and a foamed core. This structure makes it possible to achieve a larger cross section, and thus more stiffness, in the component while using the same material input and weight [2].

Because the surface quality of foam-injected pieces was insufficient for many applications, the standard method was further developed in order to generate improved surfaces. The gas counter pressure (GCP) technique prevents premature expansion of the gas by creating counter pressure by injecting gas in the tool cavity [2]. The melt can attach to the tool surfaces and develop a compact skin layer without collapsed foam pores. This greatly improves the surface quality (see Section 2.4.2). In addition to the GCP method, other methods were developed that produce components with a skin layer that consists of thermoplastic material without a blowing agent (i.e., the sandwich method) [2]. During this process, a part of the cavity is filled with melt that does not contain a blowing agent. Later, it becomes the surface of the component. Subsequently, melt containing blowing agent is injected until a certain filling degree is reached. Finally, the cavity is completely filled with blowing agent-free melt, in order to make the entire surface out of compact material. Owing to this process control, collapsed foam pores cannot attach to or freeze at the surface.

Another strategy that aims to improve the surface is the variotherm method (Section 2.4.3). This method is based on tempering the injection molding tool to prevent a freezing of the inrushing polymer-gas mixture on the tool wall. The high temperature keeps the polymer in a melted state in the skin area. This generates a homogeneous skin layer without including collapsed foam pores [4].

Microcellular foams represent a group of newer developments in the field of thermoplastic foam injection molding and display an average pore size of less than 10 μm [2]. They are manufactured using the direct gassing method, and, in addition to their pore size, they are characterized by a large number of pores (10^9 pores/cm^3) [2]. In comparison to standard foam, microcellular foams make it possible to achieve a lower decrease in the mechanical values. Also, the smaller foam pores makes it possible to produce thin-walled components, which could not be made without flaws when using standard pore sizes [2].

2.1.2 Typical Applications

If the components that are manufactured by using thermoplastic foam injection molding were not made with one of the above-mentioned techniques, they usually have the disadvantage that they cannot be used in visible areas. Postprocessing of the surface or special process control can remedy this, as mentioned previously. However, both involve additional costs due to either follow-up processing or the acquisition of special machines. Thermoplastic foam injection molded components are especially suitable for application in large or strongly ribbed injection molded pieces. Moreover, they function well in applications for which good dimensional stability is a prerequisite. Since the first use of thermoplastic foam injection molding, casing pieces for electronic devices often have been made using foam injection molding. The reason for this is that this method creates large wall thicknesses due to the foaming of the melt, and simultaneously prevents sink marks toward stiffening ribs while also guaranteeing high dimensional stability [2] (Figure 2.3).

The blowing agent in the polymer reduces the viscosity in the melt, as mentioned previously, which enables longer flow paths in the cavity. For this reason, thermoplastic foam injection molding is often used to manufacture particularly thin-walled geometries for which the melt must cover a large flow path without solidifying too early (i.e., in packaging).

The developments of the variotherm and GCP methods made it possible to employ foamed components in visible areas. The automotive

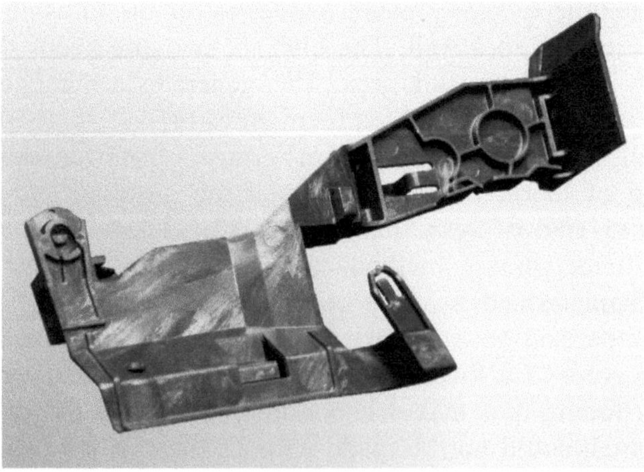

Figure 2.3 Typical foam injection molded component for use in nonvisible areas (Ford).

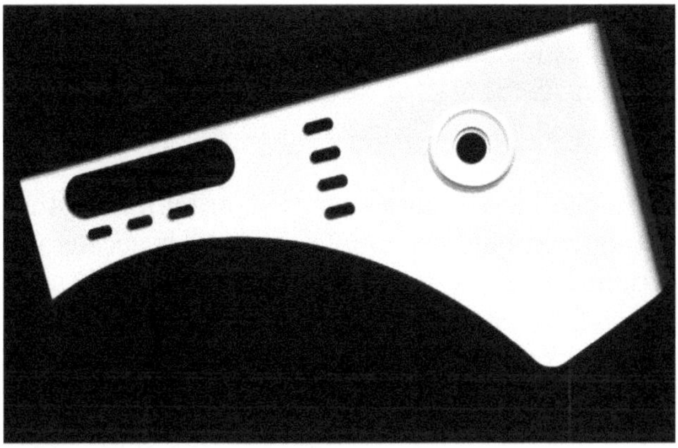

Figure 2.4 Washing machine panel (Miele) produced by Krauss Maffei.

industry already utilizes the weight reduction and high dimension accuracy benefits for individual components used in visible areas. Furthermore, thermoplastic foam injection molding is used in manufacturing for applications ranging from small fixtures to entire instrument panel carriers. In other industries, visible components are manufactured as well, i.e. a washing machine panel that is about 35% lighter than the compact piece [5] (Figure 2.4).

The manufacture of soft-touch surfaces in interior car parts is relatively new. Compact components are covered in foam in an additional processing step in order to modify the haptic of the surface. This method makes it possible to use fewer processing steps and to simplify the relatively intricate manufacture of soft-touch surfaces [6].

2.2 Foaming Mechanism

2.2.1 Solubility and Diffusion

A single-phase polymer-gas mixture is required to create a homogeneous foam structure. Its creation depends upon the technical prerequisites, as well as the solubility of the employed gas in the polymer [2].

The amount of gas that can be dissolved in a polymer varies depending on the type of polymer and blowing agents [7]. The solubility increases in correlation with the pressure and decreases with increasing temperature. The maximum sorption capacity of the polymer determines the maximal foaming degree later in the process. The dissolved gas in the polymer causes the viscosity to decrease greatly and raises the glass transition temperature slightly [2]. The gas molecules diffused between the macromolecules increase the mobility of the macromolecular chains. After the addition of the gas, it is distributed throughout the diffusion processes in the polymer melt, leading to a single-phase polymer-gas mixture [2].

In order to achieve a homogeneous distribution of the gas in the polymer, mixing elements are employed, which accelerate the homogeneous distribution of the gas in the blowing agent by shortening the diffusion paths. Both static and dynamic mixing elements are used. The higher the diffusion speed of the gas in the polymer, the less time it takes to achieve an homogeneous distribution in the melt. The diffusion speed depends on the employed gas-polymer mixture and is referenced in the research literature for CO_2 and N_2 in various polymers as less than 10 s [8].

2.2.2 Nucleation

Nucleation marks the beginning of the formation of foam pores and can take place in two ways. Heterogeneous nucleation occurs in the interfaces in the gas-polymer mixture, which are caused by filling materials, reaction residues, or impurities [9]. Homogeneous nucleation occurs when no additional materials are in the melt and hence there are no interfaces [9]. In practice, it can be assumed that a combination of homogeneous and inhomogeneous nucleation takes place. A purely homogeneous nucleation would be possible only in a homogeneous system without impurities [2]. The required nucleation energy for heterogeneous nucleation is lower than that for homogeneous nucleation, and this is the reason why heterogeneous nucleation takes place before homogeneous nucleation. Depending on the wetting angle between the polymer and foreign material, the required nucleation energy is lower if the wetting is poor [8]. Detailed information concerning the calculation of nucleation rates can be found in several publications [2,4,8,9].

2.2.3 Cell Growth

During the subsequent growth of pores, it is assumed that hydrodynamic forces exert a significant influence in the initial phase. At first, the pores grow separately from one another, so they can be regarded as an isolated system in a polymer melt [8]. Growth leads to a reduction of pressure inside the pore, causing the gas molecules dissolved in the

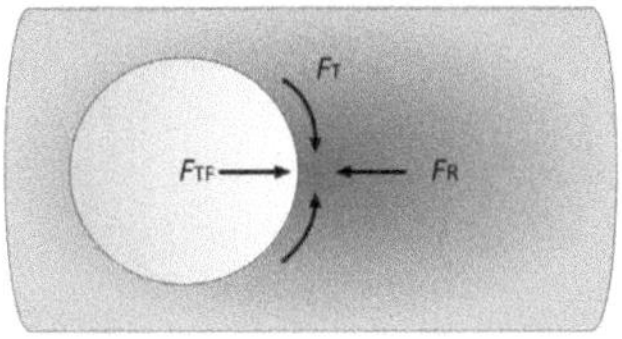

Figure 2.5 Illustration of the forces that affect a foam pore according to [10].

polymer melt to diffuse into the pores and a concentration gradient to occur in the polymer melt around the created pore [10]. The pores continue to grow until the gas concentration in the entire melt has decreased so much that no more gas can diffuse into pores. If sufficient free volume is available, the pore's growth ends when an equilibrium is reached between the force of the gas pressure and the force needed for further growth [2]; see Figure 2.5.

If the shells of two pores touch, then only the gas in the respective shell is available for growth of the pore [8]. If two differently sized foam pores come into contact, the smaller one diffuses into the larger pore due to the difference of the internal pressure of the pores. When two foam pores merge into one, this phenomenon, referred to as *coalescence*, has a negative effect on the mechanical properties of the foam injection molded part. The description of the static state of a pore illustrates the dependency of the internal pressure on the radius as follows:

$$\gamma_S \cdot dA_0 = p_B \cdot dV \tag{2.1}$$

where

γ_S = Surface tension
A_0 = Pore surface
p_B = Internal pressure of the pore
V = Pore volume

After employing the specific equation for the surface and volume of a sphere, the following equation results:

$$\gamma_S \cdot 8 \cdot \pi \cdot r \cdot dr = p_B \cdot 4 \cdot \pi \cdot r^2 \cdot dr \tag{2.2}$$

where

r = Pore radius

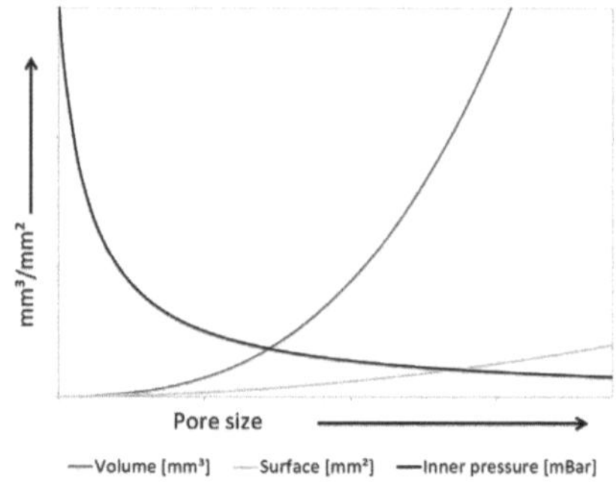

Figure 2.6 Dependency of the internal pore pressure on the volume and surface of the pore.

After converting the equation, the internal pressure of a foam pore results in dependency of the surface tension and the pore radius:

$$p_B = \frac{2\gamma_S}{r} \tag{2.3}$$

Figure 2.6 illustrates how the partial pressure inside the pore decreases as the size increases. The volume grows larger in proportion to the surface. Assuming the temperature and surface energy remain constant, Eq. (2.1) verifies that the decreasing internal pore pressure and diffusion of smaller pores into larger ones are caused by the higher pressure.

Because simultaneously developing foam pores theoretically have the same volume when they come into contact, even nucleation is of decisive significance for a homogeneous foam structure [8] in order to prevent coalescence caused by pressure differences.

During pore growth, the melt is stretched on the pore surface, which leads to an orientation of the stretching in the direction of the macromolecular chains. If the melt strength is too low, the pore walls between foam pores can tear, resulting in an unwanted foam structure [8].

The viscosity of the melt decreases with the increasing temperature because the mobility of the macromolecules increases. This gives the blowing agent shorter diffusion time frames because the gas molecules can move in between the molecular chains more easily. Owing to the enhanced diffusion speed (and thus, gas diffusing into the pores more quickly), the pores grow faster. In addition, the melt strength reduces due to the lower viscosity, which, in turn, leads to a reduction of the tangential forces of the polymer on the pore wall. This is also beneficial for pore growth. As a consequence, the pore diameter increases with the temperature. However, this also poses the threat of unwanted pore coalescence.

2.2.4 Stabilizing Effects

In order to obtain an homogeneous foam structure, the foamed melt must be stabilized at the right point in time. The viscosity of the polymer increases due to the gas diffusing out and cooling in the tool [2]. As a result, the extensional viscosity of the melt increases, as the volume work for further growth of the foam pores does. How fast the foaming polymer melt is stabilized largely depends on the wall thickness of the foam injection molded part and the tempering of the tool [2]. If the foam injection molded part is very thin and the tool temperature is low, relatively thick skin layers result, which lower the maximum achievable density reduction. In thick-walled parts and at high tool temperatures, diffusion processes can lead to unwanted pore coalescence in the core layer [2].

Another factor that must be taken into account during foaming is the low ambient pressure in the injection mold. This can allow gas from the melt to diffuse into the cavity, and these gas losses can lead to a lower foaming degree. But they can be minimized by a quick freezing of the skin layer, which in correlation leads to a reduction of the diffusion speed [8] (see Section 2.2.1).

Additionally, it is important to keep in mind that the cavity limits the volume during foam injection, which makes the stabilization of the melt and the size of the foam pores dependent on the cooling, the resulting viscosity increase in the melt, and the volume available in the cavity. The pores can only grow until the cavity is volumetrically filled, even if sufficient gas pressure is available for further expansion. A state of equilibrium occurs, and the size of the foam pores remains constant. Depending on the viscosity of the melt, it is questionable how long this state of equilibrium can be maintained. If the viscosity of the melt is too low due to a high temperature, a disintegration of gas in the polymer and a diffusing of individual pores into one another can occur. Moreover, a diffusion of the gas out of the melt, and a collection of it between the polymer and the tool wall can also result. For this reason, a stabilization of the foam pores by rapid cooling of the melt is indispensable. If cooling is sufficient, further growth of the pores when volumetrically filling the cavity is possible only if the volume of the melt shrinks during cooling, as shown in Figure 2.7.

The shrinkage is equalized by further expansion of the pores if the blowing agent pressure inside the pore is sufficient for further expansion. In this case, the blowing agent pressure must be able to guarantee further growth by supplying internal pressure to the pore walls, despite the viscosity increasing due to cooling and the resulting tangential

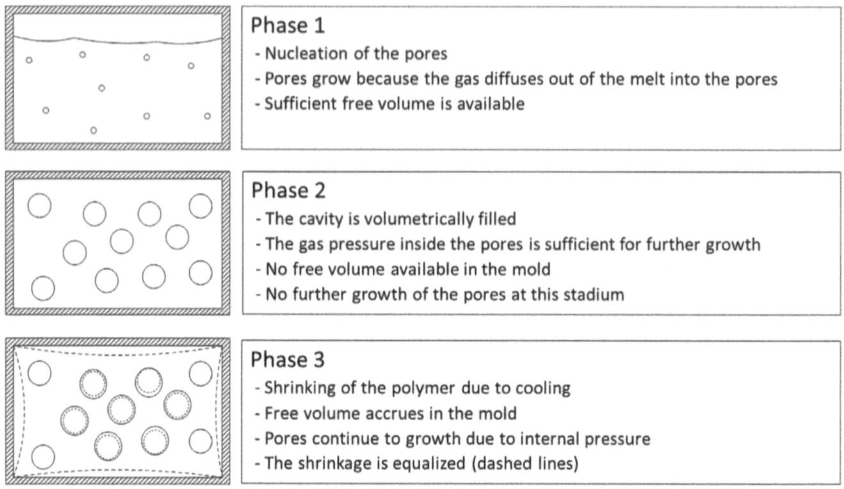

Figure 2.7 Dependency of the pore growth on the free volume in the cavity.

force. As the temperature decreases, this effect is prevented because the tangential force, which affects the pore wall, continually increases. Consequently, the blowing agent pressure cannot continue to rise once the gas from the polymer has completely diffused into the pores. This process takes place analogous to the temperature profile in the melt. Accordingly, the growth in the pores stops from the tool wall to the core. Here, it is important to continue the cooling process until the core areas of the component have cooled enough to prevent further expansion and a subsequent swelling of the components after demolding.

2.2.5 Physical and Chemical Blowing Agents

Chemical and physical blowing agents are common in injection molding. Chemical blowing agents are added in solid form and dissolve throughout the course of the process, while physical blowing agents are injected directly into the melt in liquid form. Regardless of their form, the employed blowing agents must possess high solubility when in the polymer, be easy to incorporate, enable a homogeneous distribution in the melt, and produce very small amounts of solid reaction products [11,12].

2.2.5.1 *Physical Blowing Agents*

Physical blowing agents are fluids that are added to the already-melted polymer during plasticization in the injection unit [13].

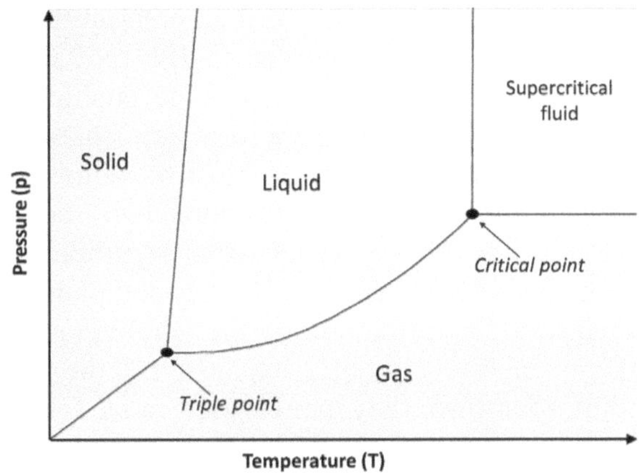

Figure 2.8 Supercritical fluid CO_2.

Foaming is induced by an expansion of the blowing agent when the aggregate state changes. Due to the high pressure and high temperature in the plasticization unit, the blowing agents are in a supercritical state when they are mixed into the polymer. They promptly expand when the melt enters the cavity because of the intense pressure drop and the subsequent transition of the blowing agent into a gaseous state. This volume increase in the polymer creates the foam structure [14,15].

When employing inert gases like nitrogen (N_2), carbon dioxide (CO_2), and hydrocarbons [13], no discolorations of the polymer occur during physical foaming, and the physical properties of the polymer remain unaffected. The opposite is the case for chemical blowing agents due to decomposition products that can result from the blowing agent reaction (Section 2.2.5.2).

In order to achieve a high degree of solubility in the polymer melt, the physical blowing agents are usually added in an overcritical state. If a fluid is in this state, liquefaction is no longer possible due to an increase in pressure, which makes addition reproducible. In an overcritical state, fluids also display low viscosity, low surface tension, and good diffusion properties, all of which are responsible for quick wetting and excellent solubility in the polymer (Figure 2.8). Moreover, they provide a high nucleation density, which helps to create an homogeneous foam structure [16].

The supercritical points of the blowing agents are:

CO_2: pC: 73.84 bar TC: 31°C
N_2 : pC: 33.90 bar TC: −147°C

The process dynamic induced by the high pressure in the plasticization unit and the high injection speeds in the tool require complex engineering systems (precise and fast machine positioning, nonreturn valves, back pressure control, precise pressure control, gas dosing, etc.). Moreover, a highly back pressure must be achieved during plasticization in order to dose the gas and homogenize the blowing agent in the polymer melt. For this purpose, special injection molding equipment is required. This technique entails substantial investment costs but also can lead to a cost reduction on the whole [17,18]. The physical blowing agents enable very quick expansion in the tool due to their low boiling point; therefore, they can be used in all foaming technologies in which an homogeneous and high density reduction is the aim [19,20] (Section 2.4.5).

2.2.5.2 Chemical Blowing Agents

Chemical blowing agents are added to the polymer either prior to or during plasticization. They consist of organic or inorganic compounds that dissociate gas due to a reaction during processing [21,22].

The group of chemical blowing agents can be categorized as exothermic and endothermic. Exothermic blowing agents release energy during a reaction, which must be dissipated by an adjusted tempering of the plasticization unit and tool. After the activation temperature is exceeded, the reaction runs its course without more energy being added, and continues until the blowing agent is completely done reacting. In the case of blowing agents that react endothermically, energy must be continuously applied in the form of heat so that the reaction does not stop. Endothermic blowing agents usually dissociate water when reacting, which can lead to a hydrolytic degradation of the polymer chains. Chemical blowing agents differ strongly with regard to the amount of activation energy needed, their reaction speed, and the amount of gas produced [23].

In technical thermoplastics, endothermic blowing agents have several technological advantages compared to exothermic blowing agents (i.e., shorter cycle times, finer pore structures, smoother surfaces, thicker skin layers, controlled temperature development, and more precise process control due to influencing the processing temperature). Disadvantages of exothermic blowing agents include a smaller temperature window in which endothermic agents can be employed, corrosion of the tool and screw in the case of unalloyed steel, and reduced gas pressure (as discussed later in this chapter) [23]. The gas yield varies strongly and determines the required dosage of the blowing agent, as well as the maximum achievable density reduction [20].

Chemical blowing agents are commonly processed with a polymeric carrier as their basis and in the form of a master batch, which guarantees that its addition to the polymer to be foamed will be a simple process. The blowing agent is, among other parameters like the compability of polymer and exothermic or endothermic reaction of blowing agent, selected on the reaction temperature. Here, the employed mass temperature of the polymer must be above this threshold value. The polymeric carrier must also be compatible with the matrix polymer and display a softening or melting range below the reaction temperature of the blowing agent. Otherwise, the production of a master batch would not be possible due to the occurring gas development.

The decomposition residues that are created during the reaction of chemical blowing agents function as nucleation agents during later foaming and promote a fine pore structure. However, they can also affect the material properties negatively (i.e., embrittlement) or lead to discoloration of the component [13,23]. Several blowing agents produce acid as a residual product of the chemical reaction. Consequently, it is necessary to use corrosion-resistant screws and plasticization cylinders in processing [2].

The benefit of chemical blowing agents is their less expensive processing technology, which only requires a plasticization unit with a shutoff nozzle, and back pressure control. Process control and flexible production are benefits as well. Especially when there are fewer requirements on the foaming degree and mechanical properties, chemical blowing agents offer technological and economic advantages in comparison to physical blowing agents.

2.2.6 Morphology of Foams

The formation of foam is a complex process because thermally and mechanically superimposed processes take place simultaneously. The moment the shutoff nozzle is opened, the injection process and the expansion of the blowing agent begin. As a consequence, the density gradient is oriented in the direction of the flow. Injection molded foam components always display an integral foam structure that can be attributed to the locally differing temperatures and the flow processes during processing. The tool wall is low tempered compared to the polymer melt, leading to an immediate freezing of the polymer when it comes into contact, and inhibit foaming. The blowing agent dissolved in the polymer remains in the component's skin layer and

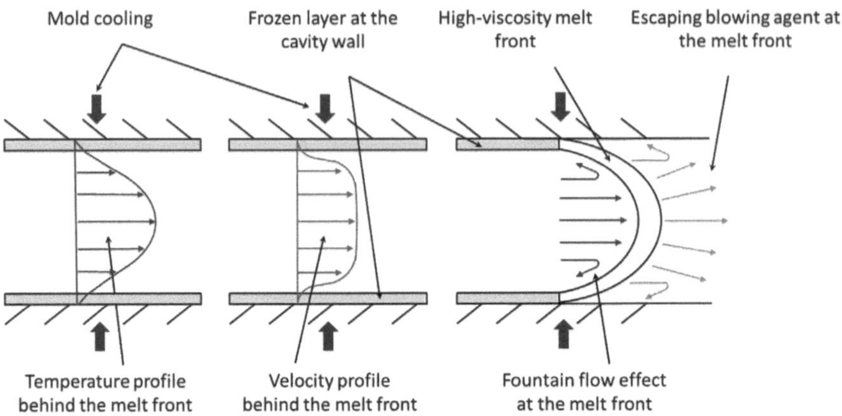

Figure 2.9 Schematic illustration of the headstream in the injection molding tool [2].

diffuses out of the material later. The freezing process on the tool wall induces a frontal laminar flow in the component's center. This, in turn, creates the typical integral foam structure of thermoplastic foam injection molding, with a compact skin layer and a foamed core (Figure 2.9) [2,3,14,24].

The morphology of plastic foams has a decisive influence on the mechanical properties. Commonly, the morphology is characterized by the pore volume ratio or the pore density (number of pores per area) and the average pore diameter [25–29]. Due to the integral structure of injection molded foams, this is only conditionally suitable for a precise characterization of the morphology.

It is common knowledge that a compact skin layer cannot be prevented in injection molded polymer foams. The resulting inhomogeneity of the density and the foam structure complicate characterization. Kirschling [24] divided the foam morphology into five sections (Figure 2.10). The division comprises the compact skin layer, which reaches from the skin layer of the sample to the first pore, and the foam core. The foam core itself is divided into one inner-core area and two outer-core areas.

In addition to the thickness of the skin layer, the average pore diameter is an important value for the characterization of the morphology of injection molded thermoplastic foams. The average pore diameter is mainly determined by the amount of blowing agent. The higher the amount of blowing agent, the larger is the nucleation in the polymer melt. When using physical blowing agents, a higher nucleation rate can create smaller pore diameters than chemical blowing agents.

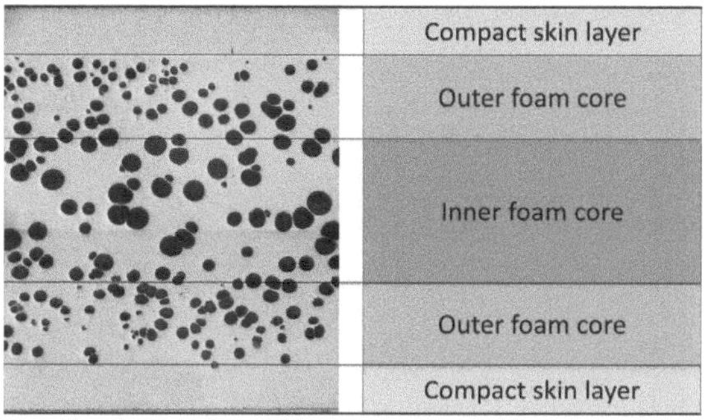

Figure 2.10 Division of injection molded foams into morphological areas (cross section) [30].

The distances between two neighboring pores represent the framework of the foam core and significantly determine the mechanical properties. While the distance between foam pores is a common parameter for characterization foams with a density around 0.04 to 0.30 g/cm^3, determining the distance between pores is uncommon for injection molded foams so far. Initial examinations of pore distances have shown that they have a definite influence on the mechanical properties of thermoplastic foams [7,24].

2.2.7 Mechanical and Optical Properties

In thermoplastic foam injection molding, foamed polymers are characterized by specifically adjustable properties and extensive application possibilities. In comparison to compact polymers, they display low heat conductivity, lower density, and better acoustic properties. In addition to the known, beneficial properties of thermoplastic foams, the mechanical properties continue to gain in importance. At first, it was assumed that stiffness and strength decrease proportional to the reduction of density. Today, it is known that the decrease of the mechanical properties is exponential. By means of optimized foam structures, the properties can be specifically optimized within physical limits. Especially when exposed to bending strains, injection molded foam pieces display great potential due to their integral density profile [31,32].

2.2.7.1 Mechanical Properties

Due to the highly complex injection molding process in combination with foaming, the process can be simulated only conditionally.

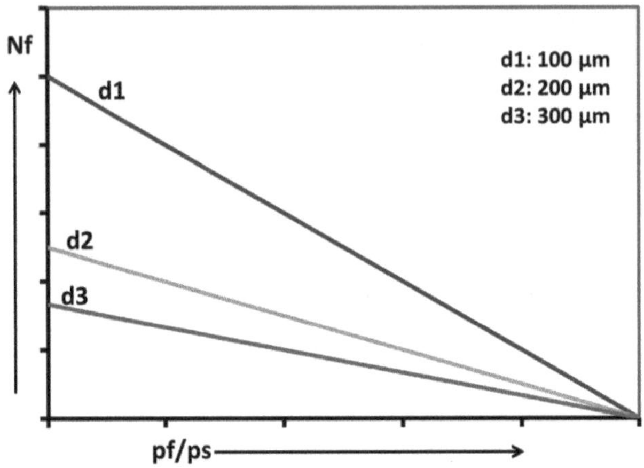

Figure 2.11 Estimated pore density (N_f) over the density ratio of foam (p_f) in relation to the compact material (p_s) for varying average pore diameters.

However, model approaches exist, which can be used to predict the component properties based on the foam structure or the process parameters (Figure 2.11).

The pore density is defined by the following equation [33]:

$$N_f = \left(\frac{n \cdot M^2}{A}\right)^{\frac{3}{2}} \tag{2.4}$$

where

N_f = Pore density [pores/cm^3]
n = Number of pores in the image
A = Area of the image
M = Magnification factor of the image

and can be estimated by the density of the component [34]:

$$N_f = \frac{1 - \frac{p_f}{p_s}}{10^{-4} \cdot d} \tag{2.5}$$

where

p_f = Density of the foam
p_s = Density of the compact material
d = Average pore diameter

Moreover, approximation equations by Klempner and Sendijarevic exist, which enable a calculation of the average pore distance and the pore volume based on the average pore diameter and the pore density [34]:

$$\text{Pore distance:} \quad \alpha = d\left(\frac{1}{\sqrt{1-\frac{\rho_f}{\rho_s}}} - 1\right) \tag{2.6}$$

$$\text{Pore volume:} \quad V_f = \left(\frac{\pi}{6}\right) \cdot D^3 \cdot N_f \tag{2.7}$$

where

α = Pore distance
d = Average pore diameter
ρ_s = Density of the compact material
ρ_f = Density of the foam

Owing to the integral foam structure, an improvement of the stiffness of components was achieved using thermoplastic foam injection molding results in relation of the weight, especially when exposed to bending strains. In order to be able to better estimate the mechanical properties of foamed components, the bending strength can be calculated approximately because there is a correlation between compact and foamed components regarding the density and E-modulus [2].

Initial approaches for an approximation of the mechanical properties by Gibson and Ashby [35] describe the correlation between a compact material and foam as follows:

$$\frac{E_f}{E_s} = \left(\frac{\rho_f}{\rho_s}\right)^x \tag{2.8}$$

where

E_s = Elasticity module of the compact material
E_f = Elasticity module of the foam
ρ_s = Density of the compact material
ρ_f = Density of the foam
x = Exponent that must be determined experimentally ($1 < x < 2$)

In Figure 2.12, real values of a foamed polycarbonate (PC) are shown that were determined by tensile testing in accordance with DIN

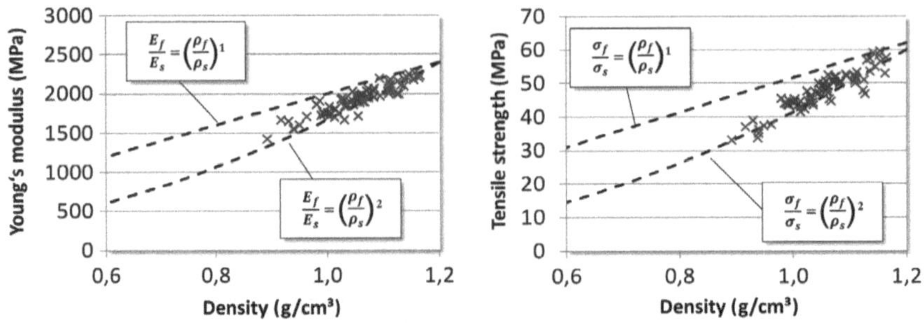

Figure 2.12 Tensile properties of the examined PC foams depending on the density (Makrolon 2805, DIN EN ISO 527) [30].

EN ISO 527. They are compared to values that were estimated using the density ratio of foam to the compact material.

The result of the calculation strongly depends on the exponent x, which reflects the correlation between the density and elasticity and, ideally, strives toward 1. In practice, common values range between 1.5 and 2. However, the foam morphology is not taken into account; therefore, this is only a rough estimation. Nonetheless, this method represents the conventional method for the estimation of the mechanical properties of polymer foams [35–38].

The first attempt to take the sandwich structure of injection molded integral foams into account was made by Müller, who divided the material into a classic sandwich in his value simulation, with a top layer with a high density and stiffness, and a homogeneous foam core with a low stiffness [39]:

$$\text{Young's modulus:} \quad E_T = E_S(1 - k_e) + E_f k_e \tag{2.9}$$

$$\text{Flexural modulus:} \quad E_B = E_s(1 - k_e^3) + E_f k_e^3 \tag{2.10}$$

where

E_S = Stiffness of the compact material
E_T = Tensile E-modulus of the foam
E_B = Flexural modulus of the foam
ρ_s = Density of the compact material
k_e = Effective ratio of the foam layer of the integral foam

The stiffness of the foam core is estimated as follows:

$$E_f = E_s(1 - \Delta\rho_s)^l. \tag{2.11}$$

The exponent depends upon the component thickness and ranges, as derived from Gibson and Ashby, between 1 and 2. However, values from 1.6 to 2 are considered more practical [39]. This approach does not take the exact foam morphology into account and is, for this reason, not precise either.

The influence of the matrix material can be problematic when calculating the component toughness because it is hard to make a universal formulation for it. In addition, there is no strong correlation to the relative density of the sample.

According to Throne [38], a simple estimation can be made as follows:

$$I_f = I_s \left(\frac{p_f}{p_s}\right)^k k^l, \tag{2.12}$$

where

I_f = Impact strength of the foam
I_s = Impact strength of the compact material
p_f = Density of the foam
p_s = Density of the compact material
k, l = Empirical coefficient

Only the density and toughness of the compact material are regarded for the calculation, while the foam structure is left out entirely. There is another approach by *Xu* for the calculation of the relative notched impact strength according to Izod in comparison to the compact material [40].

$$\text{Izod notched impact strength:} \quad R_i = \sqrt[3]{\frac{N_m}{N_s}}(1-R_n)^4\left(1-\frac{2t}{h}\right)^{-1} \tag{2.13}$$

where

R_i = Izod notched impact strength ratio between the foam and compact material
R_n = Density reduction ratio of the foam core
t = Skin layer thickness
h = Component thickness
N_s = Pore density of conventional foams, calculated with a pore diameter of 100 μm
N_m = Pore density of microcellular foam

N_m is calculated using the following equation:

$$N_m = \frac{1 - R}{0.0001\,d}\left(\frac{1}{cm^3}\right) \tag{2.14}$$

where

d = Average pore diameter in millimeters
R = Density reduction of the foam core in comparison to the compact material

Xu et al. [40] report that this equation can be used conveniently to predict the notched impact strength of polymer foams and that the results of different studies were similar to the calculated notched impact strength.

2.2.7.2 Optical Properties

One disadvantage of foam injection molded components is the substandard surface quality, as already mentioned, which is a result of the expansion of the blowing agents triggered by the pressure drop during injection [2,41]. Due to the pressure drop, an oversaturation of gas in the melt results and pores develop on the melt front. When these pores burst under shearing stress and are carried to the surface of the tool wall by the frontal laminar flow, streaks appear and the surface becomes rough. The gas trapped between the following additional melt and the tool surface gives the surface of the molded part the typical pattern, which consists of microscopic indentations on the component surface. This effect not only affects the appearance of the component but also the mechanical values. Especially in plastics that are sensitive to notching, like PC, these indentations on the surface cause notch effects.

Suitable process control makes it possible to improve the surfaces, but they remain the main reason for the still-very-limited presence of foam injection molded components on the market [2,42,43]. In addition to streaks [44], other surface defects exist that are caused by solid state deformation. They lead to higher levels of roughness, and also to a poorer surface impression [45].

This is where special processing technologies like the GCP technology, variotherm tools, and precision opening tools can help (Section 2.4). The comparison of two samples depicted in Figure 2.13 shows the different surface qualities achieved with and without the

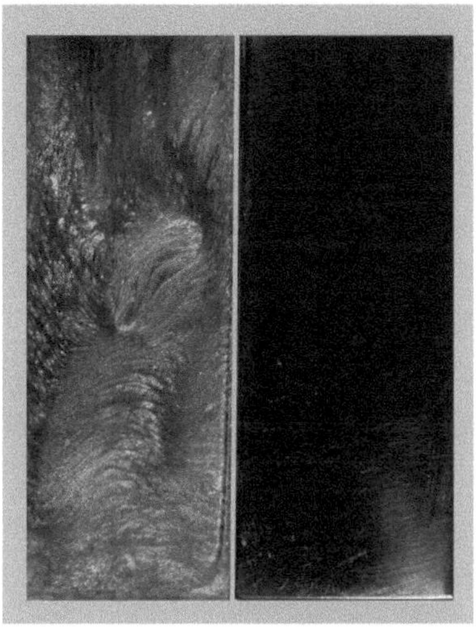

Figure 2.13 Comparison of a component made with and without the GCP technique.

GCP technology. The sample shown on the left side of the image displays the surface structure typical for thermoplastic foam injection molding, while the sample on the right shows the structure typical for the GCP technique.

2.3 Process Route

2.3.1 From Granules to Technical Parts

Thermoplastic foam injection molding is very similar to standard injection molding, which makes it possible to manufacture chemically foamed components on commercially available injection molding machines without having to exert great effort into acquiring additional equipment [46] (Figure 2.14). A large pressure drop between the plasticization unit and tool cavity induces the melt to foam. The formation of foam is achieved by the gas pressure in the polymer melt, regardless of the blowing agent employed. A single-phase polymer-gas solution is essential for reproducible and homogeneous foaming. It is created in the plasticization unit by suitable mixing elements. Two machine components are fundamental for the production of foamed injection molding components. First, a back pressure control for the plasticization unit must be available. Otherwise, the polymer-gas

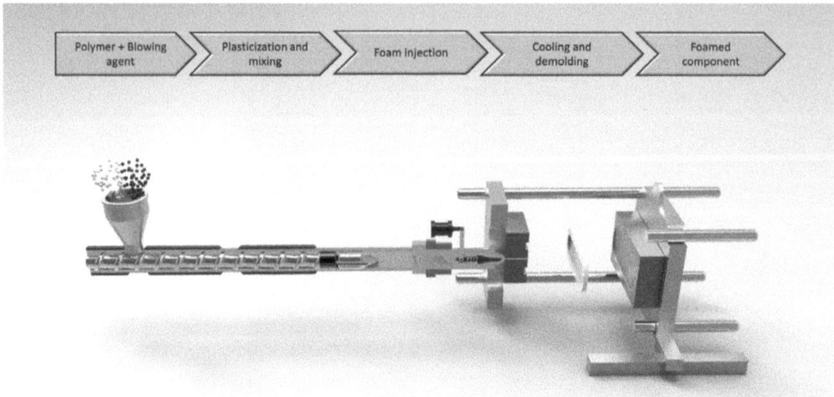

Figure 2.14 Simple description of the process chain for foam injection molding with chemical blowing agents.

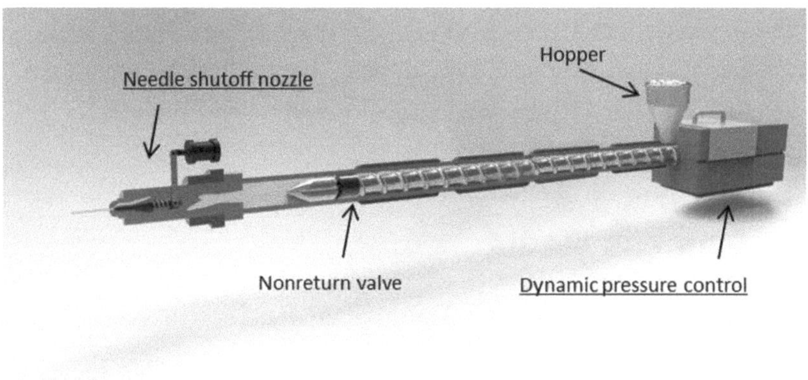

Figure 2.15 Illustration of an plasticizing unit machine with the components required for thermoplastic foam injection molding.

mixture, which is subject to pressure in the barrel in front of the screw, would move the screw axially, leading to undefined dosing of the polymer and large fluctuations in terms of the quality and weight. Moreover, the blowing agent would begin to expand in the plasticization unit and leak out during injection. Thus, no foaming would take place in the tool. The second decisive component is a needle shutoff nozzle on the plasticization unit or the hot runner system. Without it, the polymer-gas mixture would foam and leak out of the nozzle, which would also lead to a premature loss of the gas [3] (Figure 2.15).

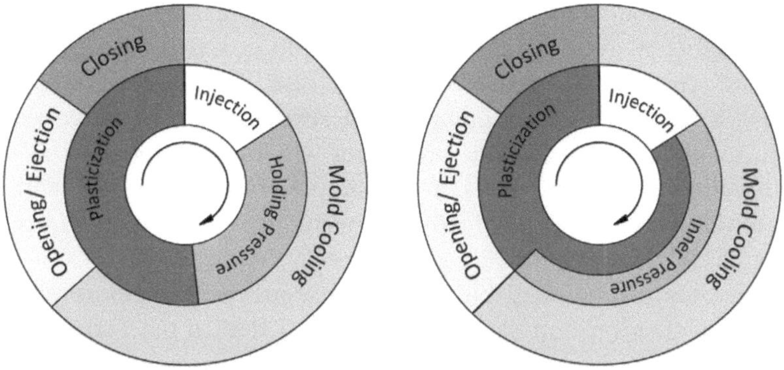

Figure 2.16 Injection molding cycle (left) in comparison to a thermoplastic foam injection molding cycle (right).

2.3.1.1 Foaming Using Chemical Blowing Agents

Adding chemical blowing agents is not complicated because the blowing agent exists as a master batch and can be mixed with the plastic prior to dosing (either manually) via a gravimetric feeder unit.

Here, there is a slight deviation from the process cycle of classic injection molding (namely, in the partial filling of the tool) and the missing holding pressure. While the tool is volumetrically filled when manufacturing compact components and shrinkage is subsequently compensated for by filling in more melt under high pressure (holding pressure phase), a partial filling of the tool is common in thermoplastic foam injection molding (Figure 2.16). This is because the volumetric shaping takes place due to the expansion of the polymer melt. Consequently, the shot volume in the plasticization unit controls the weight and the density of the foamed component at the same time. It functions as long as the pressure of the blowing agent is sufficient to fill the tool entirely [2].

Like in conventional injection molding, the polymer is melted by the plasticization unit of the injection molding machine. The required melt temperature depends on the polymer employed, the desired viscosity, and the solubility and diffusion speed of the gas in the polymer. When employing chemical blowing agents, it is important to exceed the reaction temperature to guarantee the incurrence of gas. In general, blowing agents that have a reaction range at least 10°C lower than the processing temperature of the employed polymer should be selected [2]. The added amount of blowing agent depends on the solubility of the gas in the polymer and the desired foaming degree. Common amount of blowing agent in injection molding lies between 1 and 5wt. % master batch. On the one hand, this is due to economic reasons because a higher blowing agent ratio leads to an increase in costs. On the other hand, there are mechanical reasons because the ratio of decomposition residues also increases. Decomposition residues can negatively affect the mechanical and optical properties of the component as their concentration increases. Thus, for example, blowing agents with low concentrations can be used to solely reduce the indentations (see Section 2.2.7).

In the case of chemical blowing agents, precise temperature control of the plasticization unit must be maintained due to the thermally activated reaction of the blowing agent. Especially when dealing with exothermic blowing agents, it is imperative to make sure that the reaction does not lead to a high temperature increase in the melt, in order to prevent thermal damage of the matrix material, the decomposition residues, or the filling materials. Manufacturers of chemical blowing agents have the ability to tailor their master batches to the necessary requirements by selecting the type of employed blowing agent, its concentration, additives, and the polymeric carrier (see Section 2.2.5).

Once the gas has been released, a homogeneous, single-phase polymer-gas mixture must be created. It is a crucial basic requirement for a homogeneous foam structure. High diffusion speeds and the shortest diffusion paths possible are beneficial for the distribution of the gas in the polymer. They are created by mixing the melt with suitable mixing elements. In order to improve the foam's homogeneity, rhombus or Z-mixers are attached to the end of the screw in the nozzle or, rarely, to static mixers between the plasticization unit and the tool [2,3,47,48] (Figure 2.17).

The injection of the gas-loaded polymer melt is carried out using a high injection speed to avoid premature expansion of the gas. Here, the

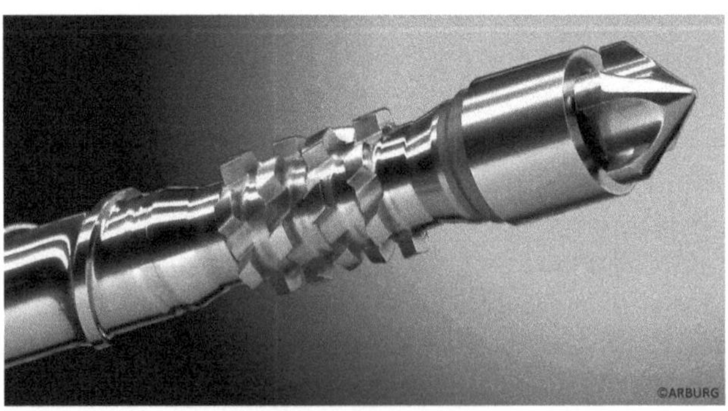

Figure 2.17 Dynamic mixing elements for injection molding (picture: ARBURG©).

filling of the cavity is not volumetric and is controlled by the shot weight or the dosing volume in the plasticization unit. However, maximum weight reduction is achieved when the maximum possible amount of gas is released, and the shot weight is sufficient to volumetrically fill the form after foaming. Because the expanding gas in the melt acts like an internal holding pressure, there is no need for a holding pressure phase when manufacturing using thermoplastic foam injection molding (see Section 2.2).

The heat that must be discharged, and, in many cases, the resulting cooling time are lower, analogous to the smaller mass of the components. However, in comparison to compact molded pieces, the cooling time can also increase, especially when paired with the poor heat conductivity of plastic foams while manufacturing thick-walled, molded pieces. It is important to make sure that the molding temperature in the core is low enough to avoid subsequent foaming of the components [49]. Furthermore, the viscosity of the polymer melt is significantly reduced by the dissolved blowing agent; thus, it also enables the production of molded pieces with very thin walls and long flow paths in the cavity. This effect is far more pronounced in physical blowing agents due to the higher blowing agent concentration and the lack of blowing agent reaction products (see Section 2.2.1).

2.3.1.2 Foaming Using Physical Blowing Agents

The requirements for the injection molding machine are much higher when employing physical blowing agents. Because the blowing agent is added to the melted polymer in supercritical state under high pressure, a gas-dosing station and a precisely controllable injection nozzle for dosing are needed. Retrofitting solutions are hard to realize, especially because additions made to the plasticization unit must be precisely timed to the process. Moreover, special screws are generally used that contribute an additional mixing in the gas-dosing area. Owing to the extra space needed for mixing elements, the length of the screw (a larger L/D ratio) changes [2,24,50,51].

2.3.1.3 Static Mixing Elements for Distributing Gas in the Melt

In addition to dynamic mixing elements, static mixing elements can be used to homogenize polymer-gas mixtures. Static mixtures do not have moving parts. Instead, they create local speed reduction by separating and rearranging the volume flow, thus creating an extensional

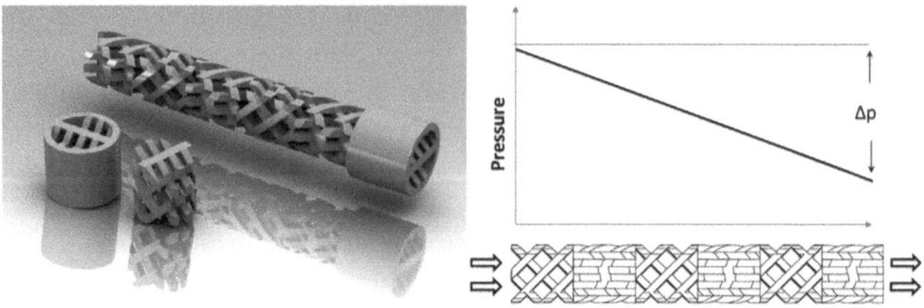

Figure 2.18 Static mixer (left) and pressure decline in a static mixture (right).

flow and a shear zone [52,53]. This type of mixer is especially suitable for physical blowing agents and makes it possible to mix the blowing agent into the melt outside the plasticization unit without needing a dynamic mixer inside the plasticization unit (see Section 2.4.5).

The increased shear stress on the polymer and the decomposition of reaction residues contained in chemical blowing agents can affect the mechanical, visual, and olfactory properties. The additional shear stress can lead to a decomposition of the reaction residues and create additional challenges for thermal management of the process, especially in correlation with exothermic blowing agents (see Section 2.2.5.2).

Static mixers are often employed for homogenization when mixing pigments in a master batch. The more homogeneous distribution can enable master batch savings [54]. Homogenization of the melt includes the added pigments and filling materials, as well as the temperature distribution [54]. Logically, the additional homogenization achieved using a static mixer can be expected to have positive effects on the foam structure because the smooth conditions cause homogeneous nucleation in the influx of the melt, which results in a homogeneous foam structure. The pressure loss, which is induced by friction in the static mixture, significantly reduces the obtainable injection speed [53,55], and lowers the pressure decline rate when the melt enters the cavity. Figure 2.18 provides an exemplary illustration of this in a static mixer. The created pressure drop can be partially compensated for by the machine if the performance limitations of the machine and mixer and the thermal capacity of the polymer-blowing agent mixture are not exceeded.

2.3.1.4 Standard Tool Technique

When designing tools like those employed to manufacture foamed components, there are special details to take into account for their designs. For example, attribute to the internal holding pressure, there is no need to worry about sink marks. Thus, it is possible to achieve higher wall thicknesses and there is more design freedom regarding the construction. Also, abrupt changes in wall thickness are a possible consequence of the internal holding pressure. These changes can lead to problems when filling the cavity. Thus, sudden wall thickness changes should be equipped with large radii [2].

Regarding mechanical aspects, designing higher stiffening ribs is also advantageous because this involves a large increase in stiffness. Sufficient ventilation in the stiffening rib for the prevention of the development of air pockets makes it possible for the melt to expand due to the blowing agent pressure, and it leads to a complete filling [2].

The low viscosity of the polymer-gas mixture enables longer flow paths and the production of thin-walled, molded pieces. In the case of these molded pieces, the maximum achievable density reduction is lower because the fast cooling of the polymer melt in the area of the tool surface inhibits the quicker expansion of the blowing agent (see Section 2.2.3).

As a rule, all components foamed using blowing agents possess a higher form precision and less warpage and postshrinkage owing to the blowing agent loading. An essential difference with the injection molding of compact materials is the positioning of the sprue. In conventional injection molded parts, it is set for the highest wall thickness. The exact opposite is the case for thermoplastic foam injection molding; the sprue is set for the lowest wall thickness [2]. Because the last portion of filling in thermoplastic foam injection molding is achieved by the internal holding pressure in the melt, and also that it is often insufficient to fill out the thin-walled areas, incorrect positioning can lead to incomplete filling of the tool or a generally lower weight reduction in the component [2]. In Figure 2.19, examples of two molded parts made using thermoplastic foam injection molding and compact injection molding are shown.

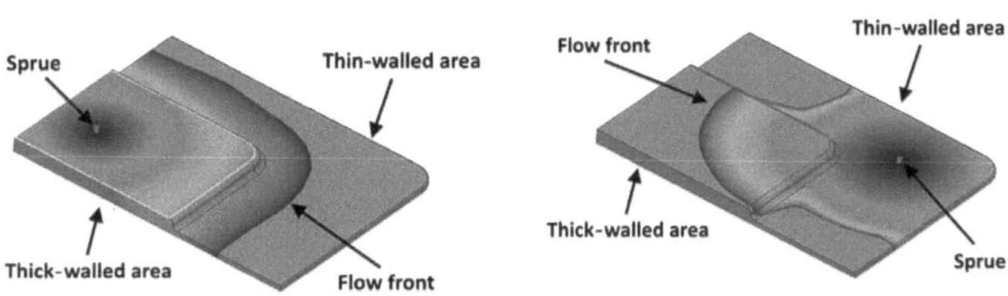

Figure 2.19 Illustration of thermoplastic foam injection molded parts with marked sprues as examples for sprue positioning for foam injection (right) and compact injection molding (left). The positions of the sprue points have been labeled with arrows.

2.3.2 Most Relevant Process Parameters

Thermoplastic foam injection molding is a highly dynamic process that strongly influences the material properties of the thermoplastics. Thus, warpage and internal stresses can be significantly controlled and optimized by the molding process [3]. As a result of superimposing the injection and foam processes, the influence of the process factors on the developing foam structure during thermoplastic foam injection molding is very pronounced. All thermoplastic foam injection molded components display the integral foam structure; however, the skin layer thickness, the pore sizes, and the pore distances can vary greatly. In the ensuing section, the most important process parameters and their influence on the foam morphology will be introduced.

2.3.2.1 *Types of Blowing Agents*

The type of blowing agent significantly influences the foam structure in the later component. Even if this is not a process parameter of the machine, the selection of the type of blowing agent is essential for process control because it interacts with other parameters. The fundamental pros and cons of chemical blowing agents have already been elaborated in detail in Section 2.2.5. They differ in terms of their chemical composition, as well as with regard to their ratio of effective components and the employed polymeric carrier.

2.3.2.1.1 Chemical Blowing Agent

The chemical blowing agent is selected in accordance with the type of plastic. This pertains to the decomposition temperature, which must match the processing temperature of the polymer, and the created reaction products (e.g., polymers sensitive to hydrolysis cannot contain water) [2]. The decomposition residues of the chemical blowing agents cause nucleation. Different sizes of decomposition residues can also help to achieve significant improvements of the foam morphology. If the processing temperatures are too high, a thermal decomposition of the reaction by-products can occur (degradation of organic hydrocarbon compounds), which is especially problematic for technical thermoplastics like polyamide (PA) and PC. It is important to take into account that foreign polymeric carriers (usually polyolefin) must be used for the master batch production due to the high processing temperature [24,56]. In the case of technical thermoplastics, primarily endothermic citric

Table 2.1 Average Pore Diameter and Skin Layer Thicknesses of Polycarbonate Foams with 2% Blowing Agent and a Weight Reduction of 10% (4 mm thickness, Makrolon 2805) [60]

Blowing agent	Polymeric carrier	Pore diameter	Skin layer thickness
Safoam RPC-40	EEA	54 ± 25 μm	1400 ± 50 μm
Hydrocerol 592	PE	73 ± 41 μm	1200 ± 50 μm
Hydrocerol 583	PS	217 ± 62 μm	1200 ± 70 μm

acid derivatives are used because azodicarbonamides, which are common for polyolefin, have a reaction temperature that is too low [57]. Several exothermic blowing agents exist that have a high reaction temperature and achieve high foaming degrees. However, various investigations have shown [58,59] that they have an evidently lower reproducibility than the conventional citric acid derivatives, and they tend to cause pore coalescences and large pore diameters. Thus, they are only suitable for producing foam structures with a high weight reduction, not microfoams. The chemical composition, the processing temperature, the additives, and the polymeric carrier (see Section 2.2.5.2) greatly influence the pore sizes. Table 2.1 illustrates this for three chemical blowing agents, using PC as the example.

The selection of a carrier material is often easy for polyolefin and other thermoplastics with low melting temperatures because a polymer that is akin to the thermoplastic can be used [61,62]. In the case of thermoplastics like PA and PC, which have high processing temperatures, this is not possible. Using polystyrene (PS) as a carrier material does not have a serious influence on the foam morphology of PC. Yet, using polyethylene (PE) and polypropylene (PP) leads to microcavities, which are caused by greater shrinking during cooling and decisively influence the morphology [56,63].

2.3.2.1.2 Physical Blowing Agents

Nitrogen gas (N_2) and CO_2 are commonly used physical blowing agents in injection molding. The gas solubility and diffusion speed of CO_2 in polymers are significantly higher than those of N_2. Owing to these properties, the blowing agent can be mixed into the plastic quicker, and higher foaming rates are achieved [64–66]. Because CO_2 nucleates faster than N_2 during injection and the melt strength is lower in this stage, CO_2 produces a larger average pore diameter. The maximum gas saturation of the melt has not been examined exactly, and results obtained using autoclave cannot be transferred to the injection molding process [65,67].

2.3.2.2 *Blowing Agent Concentration*

The following applies with certain restrictions:

"The larger the amount of blowing agent added, the larger the achievable density reduction."

Thus, the concentration of blowing agent influences the maximum weight reduction and the foam morphology. The type of blowing agent also has great significance because a larger amount of blowing agent is utilizable during physical foaming, and therefore increases the maximum weight reduction obtainable. Owing to the increasing ratio of decomposition residual products, the maximum addition of chemical blowing agents is limited. Furthermore, they increasingly influence the material properties the larger their amount and cause low reactivity if the blowing agent ratio is too high [57]. In contrast, the amount of physical blowing agent employed is only limited by the solubility of the fluid, which depends on the employed polymer, the fluid itself, and the pressure and temperature (see Section 2.2.1).

Measurements by van Krevelen [68] showed a 64-fold higher solubility of CO_2 in polyolefin than N_2. The maximum density reduction, represented as the density ratio π_ρ in Figure 2.20, when using different blowing agents cannot be equated with the blowing agent amount. Although the solubility of CO_2 in PP is clearly higher, a lower weight reduction results. The reason for this is the diffusion coefficient of CO_2

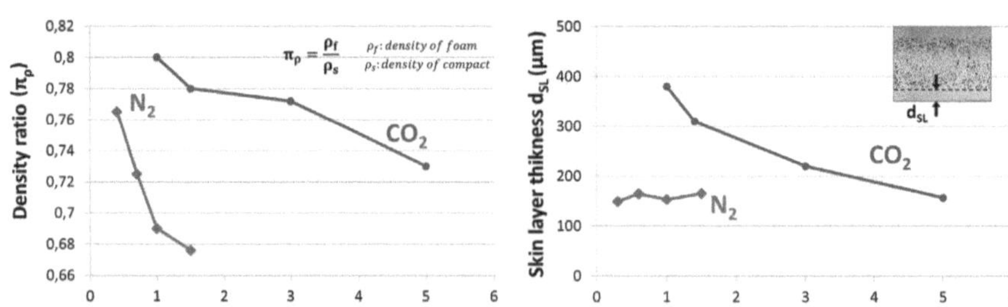

Figure 2.20 (A) Density ratio π_ρ of PP foams depending on the blowing agent and its concentration, (B) the skin layer thickness in correlation with the blowing agent and the concentration throughout a constant density reduction of 15% [2].

(it defines the speed at which the gas is mixed in). It is threefold higher in PP and PC than the diffusion coefficient of nitrogen and causes the gas to diffuse from the melt much faster. Examinations have shown that a PP copolymer displays higher N_2 solubility (1.8%) than a homopolymer (1.4%) or a long-chain, branched PP (1.2%) during foaming using a MuCell® process [7]. When using CO_2, the skin layer thickness and the pore diameter are larger than for nitrogen (Figure 2.20). This is a result of the higher diffusion speed. During injection, the gas leaks out of the polymer, and, subsequently, out of the cavity. As a result, the skin areas of the component do not foam due to the low internal gas pressure. Therefore, the sorption behavior of the polymer is a decisive factor for a foam structure with fine pores because a high blowing agent concentration causes a higher nucleation effect, which leads to a greater pore density [68,69]. Nevertheless a unlimited increase of the blowing agent ratio is not possible. Otherwise, a homogeneous distribution of the gas molecules will not result in the polymer melt, as shown in Figure 2.21. Supersaturation leads to a stagnant (or even regressive) nucleation rate, gas agglomerates, and an impaired surface quality [14,66,70]. In PC foams, the strong nucleation of a high blowing agent concentration leads to less skin layer thickness and low average pore diameter, but also increased pore distance. Thus, the foam has finer pores [24].

2.3.2.3 Mass Temperature

The employed melting temperature decidedly influences the maximum injection speed and required injection pressure. Because it essentially determines the viscosity and melt strength of a polymer, the mass temperature is of great significance for foaming. It interacts with the solubility of the gas in the polymer and the diffusion speed of the gas, which leads to a higher nucleation rate and faster diffusion of the gas out of the melt (see Section 2.2.1).

The lower the viscosity of the polymer, the less pressure needed during injection into the cavity. Low viscosity poses the risk of pore coalescence during foaming, as well as an accompanying reduction of the mechanical properties [2].

The high mass temperature delays the freezing of the skin areas, minimizing the skin layer thickness and maximizing the density reduction. A higher mass temperature results in a larger midpore diameter for PC, polycarbonate and acrylonitrile butadiene styrene (PC/ABS), and PP foams [7,24].

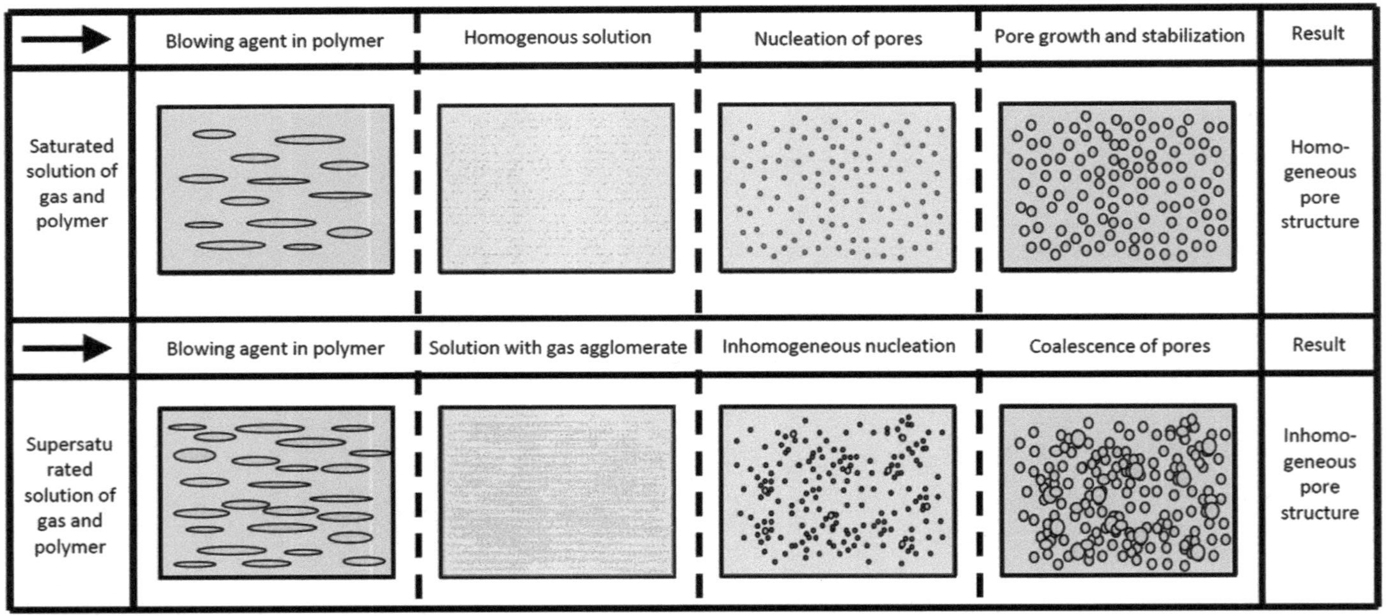

Figure 2.21 Schematically illustration of solved gas in polymer melt with resulting pore structure for saturated and supersaturated polymer gas solution.

2.3.2.4 Mold Temperature

Cooling in the melt can be controlled by the tool temperature, which also influences the developing foam morphology. A low tool temperature leads to quick freezing of the skin layer areas, increasing the skin layer thickness and slowing the diffusion of the gas into the cavity, as mentioned previously. If the tool temperature is set too low, the blowing agent expansion is prevented and the density reduction is limited. Otherwise, if the temperature is too high, the gas diffuses out between the tool wall and component surface and causes indents in the component surface [71].

Strong reciprocal effects exist between the tool temperature and melt temperature. For this reason, a high melt temperature can slightly compensate for a low tool temperature and prevent pore coalescence caused by a melt strength that is too low. In addition to a visibly larger skin layer thickness, reducing the tool temperature generally leads to a finer, more homogeneous foam structure [72–75].

Similar to injection molding compact-molded parts, the temperature of the tool crucially determines the cycle time. If the tool temperature is too high, or the cooling time too short, the material can blister after tool opening. Blisters occur when the skin layer is already frozen but the melt temperature in the core of the component is not low enough to prevent another cell increase. A high melt temperature, as a high wall thickness of the injection molded part, and an overly high blowing agent concentration promote this phenomenon. Figure 2.22 shows the temperature profile of polymer over the cavity thickness after different cooling times.

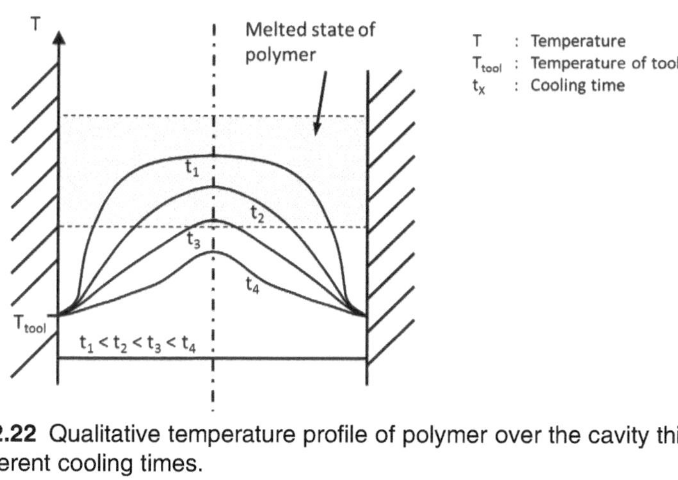

Figure 2.22 Qualitative temperature profile of polymer over the cavity thickness after different cooling times.

In the case of PC and PC/ABS foams, a reduced tool temperature leads to an increase of skin layer thickness and only conditionally influences the pore diameter and pore distances in the core [7,24,76]. To calculate the necessary cooling time for an injection molded plate-type component, the following equation can be used:

$$t_C = \frac{H^2}{\pi^2 a} \ln\left(\frac{8}{\pi^2}\frac{T_M - T_{tool}}{\overline{T}_E - T_{tool}}\right) \tag{2.15}$$

where

t_C = Cooling time
H = Component thickness
a = Thermal conductance
T_M = Mass temperature
T_{tool} = Tool temperature
$\overline{T}_E$ = Mid-demolding temperature

2.3.2.5 *Injection Speed*

The forming foam structure depends upon the pressure decline rate in the tool because the nucleation rate is higher when the pressure decline is higher. Thus, a finer, more homogeneous pore structure develops. In addition to the length of the sprue canal, the injection pressure and injection speed are mainly responsible for the pressure decline rate. Ideally, most of the expansion takes place after the injection process. This can only be achieved by a high injection speed because it leads to a rapid mold filling and achieves more homogeneous pore growth. Therefore, injection molding machines used for foaming thermoplastics are often equipped with special plasticization units, which enable an exceedingly high injection speed. Due to this high speed, the shearing near the tool wall and the orientation of the pores in the direction of the flow increase significantly [2,76].

2.3.2.6 *Back Pressure*

In order to create a single-phase gas-polymer mixture and prevent expansion of the blowing agent in the plasticization unit, utilization of back pressure is indispensable in thermoplastic foam injection molding. The required back pressure must be larger than the pressure induced by the blowing agent in the melt, and, thus, must be linked to the blowing agent concentration. If expansion of the blowing agent is not prevented prior to or during injection, coalescences occur which result in weak spots in the component. However, there is a threshold above which further increases in the back pressure cause no change in the developing foam morphology. Common back pressure values in the front of the nonreturn valve during the plasticization phase range from 100 to 300 bars, depending on the blowing agent and polymer. Physically foamed PC requires a relatively high back pressure of 200–260 bars [2,24]. In foam injection molding, a decompression lift is not used prior to or after plasticization.

2.4 Variants of Foaming Techniques

Because the process variables of injection molding machines are often insufficient to produce the quality required for thermoplastic foam injection molding components, numerous tool concepts were developed that implement additional variables in the thermoplastic foam injection molding process to vary specific properties.

Essentially, a differentiation is made between methods that prevent or eliminate foaming during injection and methods that slow the freezing of the melt by temporarily increasing the tool temperature [15,77,78]. To give an overview, Figure 2.23 shows these different techniques, which will be explained next.

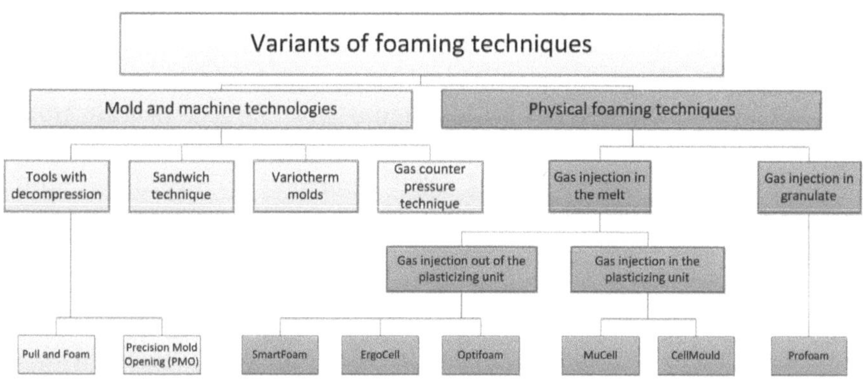

Figure 2.23 Illustration of the different foaming techniques presented in this chapter.

2.4.1 Tools with Decompression

Tools with decompression represent one form of tool technology which can achieve a targeted pressure drop by a variable volume. The technology used to manufacture tools with variable mold volumes is already known from injection compression molding and can be used throughout the entire component thickness or in sections. What is different from injection compression molding is the direction of movement of the tool, as in injection compression molding, a reduction of the mold volume takes place during the process. Otherwise, the functional mechanisms of the decompression technology are almost identical [2].

2.4.1.1 *Precision Mold Opening (PMO)*

Precision mold opening (PMO) is also referred to as a "breathing tool," "negative compression molding," "airing," or "core back expansion" [2,17,24]. PMO molds are volumetrically filled and initiate an enlargement of the cavity volume after a predefined delay. The delay time determines the skin layer thickness on the tool wall. The thickness increases proportionate to the duration of the delay time. A holding pressure is used to prevent a foaming of the polymer during the development of the compact skin layer and before the precision mold is opened. Once the delay time has passed, the cavity is enlarged, inducing an even pressure decline throughout the entire cavity and enabling the melt to foam [2,17,45,73].

Precision mold tools are used for such enlargements of the cavity volume. An accompanying position control and parallel control of the tools are necessary to prevent uneven foaming or tilting [17,18,79].

Components made using PMO possess an even density gradient across the flow path owing to the volumetric filling and subsequent expansion. Moreover, they can be manufactured with a significantly greater density reduction [44,80]. In Figure 2.24, the type of form filling and the resulting morphology of the foamed component are depicted.

By combining PMO with the GCP method, very fine-pored structures can be created. For example, a pore diameter below 10 μm is possible for PC. This results from the abrupt and homogeneous pressure drop in the melt. The opening speed of the tool and the pressure decline rate define the number of nuclei. This pressure drop causes the polymer melt to oversaturate with the blowing agent dissolved in it, and then to expand. Thus, opening the tool using a defined speed

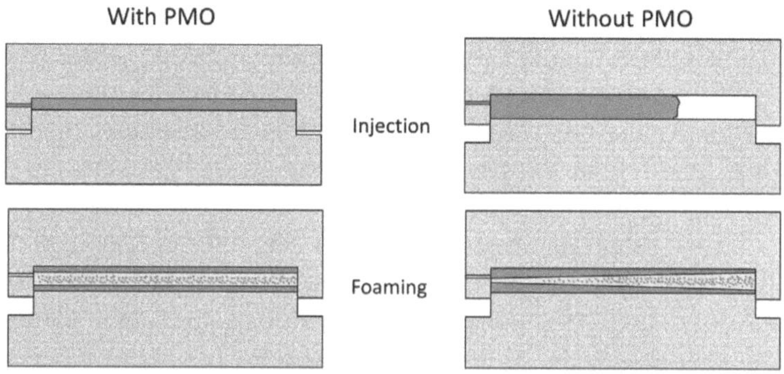

Figure 2.24 Comparison of the foam formation using a precision mold and a conventional injection molding tool [56].

profile is recommended. The maximum density reduction depends upon the remaining thickness of the plastic core and the melt viscosity of the polymer. Therefore, it is temperature dependent [81]. Due to pore coalescences, the manufacture of especially fine foam structures only works to a certain density reduction limit. Once it is exceeded, only rougher final foam structures result [2,18].

Tools with a decompression opening have essential advantages in comparison to conventional tools because they achieve a higher weight reduction and a reduced density gradient in the flow direction. Also, they can create the higher pressure decline rate needed for a fine-pored structure (see Section 2.2.2).

The two decisive parameters for the decompression opening are the opening speed, which should be quite high to increase the pressure decline rate, and the opening delay time, which plays a crucial role in the skin layer thickness and the pore diameter. It is important for the polymer melt to remain in contact with the tool surface at all times, despite high opening speeds, in order to guarantee constant and homogeneous cooling.

Changes in the cooling behavior and correlating viscosity and melt strength changes can be attributed to the influence of opening delay. The opening delay is usually set between 0 and 10 s. Negative opening delay times are possible. In this case, the tool volume already begins to expand during the injection process; this is done to achieve very high density reduction.

The skin layer thickness increases with the duration of the opening delay. However, the maximum density reduction decreases because the polymer increasingly freezes on the tool wall. Depending on the moment the tool is opened and the temperature in the core at this time, the pore diameters are affected differently. In correlation with the increased viscosity caused by the decreasing temperature, the pore diameter becomes smaller if the gas pressure is sufficient to foam the pores. If the viscosity is too high due to a low temperature, or if the gas pressure is insufficient to foam the pores, the meld is not able to expand (see Section 2.2.4).

The optimal time is always process-dependent because the opening delay time depends on the employed material, the temperature in the mass and tool, the blowing agent concentration, and the density reduction (see Section 2.3.2).

Due to the clearly lower blowing agent pressure, delay times of 0–2 s are often selected for 4-mm-thick, chemically foamed components. In the case of physically foamed components, the delay time often ranges between 2 and 8 s. In general, the larger the component and the lower the density reduction, the longer the opening delay time can be [24,57,73].

2.4.1.2 Pull and Foam

The pull-and-foam method is a special variant of the decompression method. In contrast to classic PMO, only a partial enlargement of the cavity volume is performed to manufacture specifically graded materials. Partial cavity enlargement enables the production of components with different wall thicknesses, for example stiffening ribs. After the cavity has been filled volumetrically, a high degree of foaming is achieved by enlarging the cavity volume, as shown in Figure 2.25 [82].

The principle behind this method is very similar to that of PMO and includes a volumetric filling of the cavity and an local enlargement of the cavity volume. The result is a component with hardly foamed (or even unfoamed), thin-walled areas and defined foamed thick-walled areas [83]. The expansion of the thick-walled areas allows an increase of flexural rigidity, while the mass of components remains constant. Figure 2.26 shows the flexural modulus and flexural rigidity over the density reduction and expansion rate [84].

2.4.2 GCP Method

The GCP method prevents the expansion of the gas during injection by means of high internal pressure in the tool cavity. Thus, the GCP in the cavity must be higher than the pressure induced in the melt by the blowing agent. This makes an additional sealing of the cavity necessary

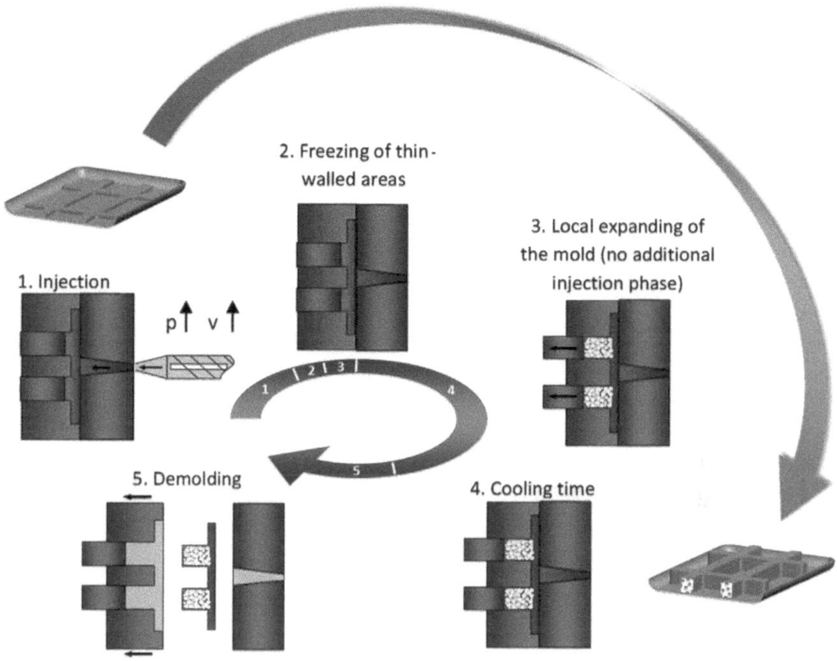

Figure 2.25 Processing sequence of the pull and foam method [83].

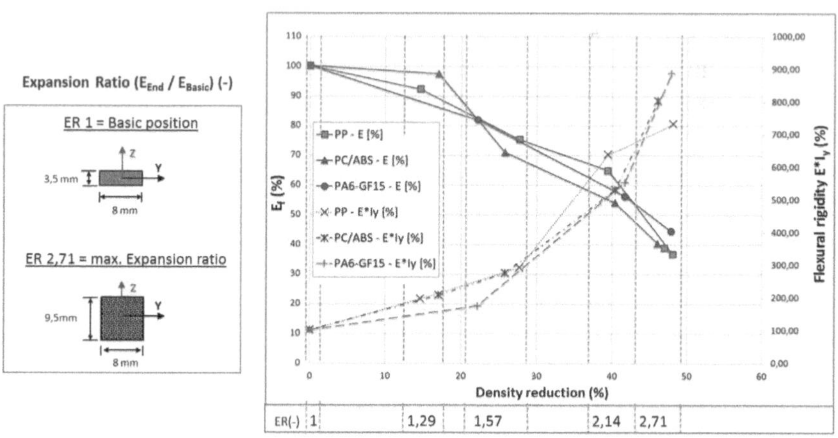

Figure 2.26 Flexural modulus and flexural rigidity in correlation with density reduction and expansion ratio [84].

[15]. The internal pressure is built up prior to the injection process, making the melt containing gas work against the gas pressure in the tool during injection and push the gas out of the tool [2,71]. A further increase of the GCP beyond that of the melt does not result in traceable

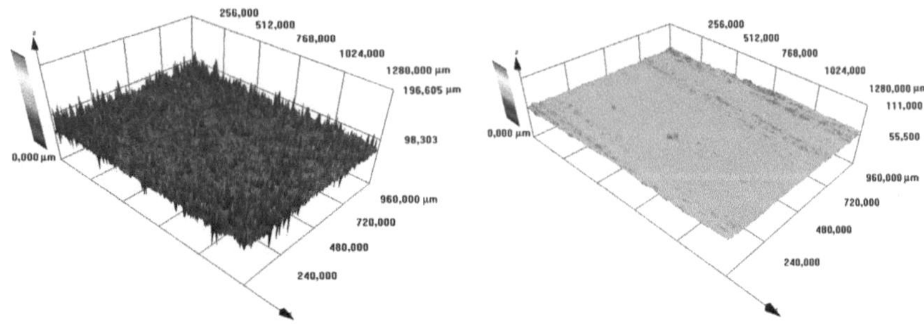

Figure 2.27 Surface roughness of a conventionally foamed PC plate (RZ = 12.3 μm) (left) and a PC plate manufactured using the GCP method (RZ = 0.4 μm) (right) (MuCell®) [30].

changes in the later foam morphology. If the GCP is less than the pressure in the polymer melt, the known and substandard surfaces occur. Due to the high costs involved in sealing the tool and the operating costs for the GCP method, it is not economical for many applications. [2,7,24,76]. Primarily, nitrogen is selected in order to avoid oxidation, chemical reactions, or impurities in the melt [18].

Injection of the gas-loaded melt against the gas pressure significantly improves the surfaces, which even makes it possible to achieve a class A status [45,85]. Figure 2.27 shows two picture analyses of surface roughness, made with a confocal laser-scanning microscope. The difference between the surface roughness of a conventionally foamed sample and a sample manufactured using the GCP method is evident.

The GCP method affects the surface quality and the morphology of the foam by means of a high pressure decline rate. An effective prevention of foaming during injection results in improved surfaces and a high pressure decline rate at the end of the filling process. The GCP method should be used in combination with a breathing tool. Alternatively, it should be used in the case of very low density reductions because the mold has to be completely filled. By partial mold filling, as it is done for conventional foam injection molds, GCP causes a strong density gradient and a rough surface quality at the end of the flow. A reduction of the injection speed is possible when using this method, which, in turn, positively affects the surface, the shearing, and the orientation of the pores. Combining the GCP with a decompression tool results in a visible reduction of the pore size, and an increase of the pore density, both of which result from the large pressure drop. In addition to a significant increase in surface quality, the GCP technique offers the possibility of

Table 2.2 Qualitative Tendencies of the Influence of the Parameters on the Mechanical Properties of PC Foams [30]

Value	Concentration of Blowing Agents	Mass Temperature	Mold Temperature	PMO	GCP	Injection Speed	Lift Delay Time
E-Modulus	↑		↓	↓		↓	↑
Tensile Strength	↑		↓	↓			↑
Elongation at Break	↑	↑	↓	↓			
Breaking Tension	↓		↓	↓	↑	↓	↑
Flexural Modulus	↓	↑	↓	↓	↑	↓	↑
Flexural Strength	↓		↓	↓	↑	↓	

influencing the foam morphology specifically by means of the GCP, holding duration, and mold opening parameters [71].

Until now, the method was almost solely used to create better surfaces, and rarely to control the morphology. For this reason, correlations between selected parameters and the resulting morphology have mostly remained unused.

In his studies, Kirschling [15,18,24,71] showed that an increase in certain mechanical properties, that is, the cryogenic temperature impact strength, is achievable by influencing the morphology.

Disadvantages of the GCP technique include extended cooling and diffusion times (the gas cannot diffuse out of the skin layers into the cavity), and a somewhat lower maximum density reduction. Even though the non-expanding gas diffuses out of the component in a matter of days, the directly ensuing thermal processing methods or coatings are a production challenge. The GCP method is mainly employed for visible parts, like casing for electronics and in the automotive industry [86].

Table 2.2 gives a summary of qualitative tendencies of the influence of the parameters on the mechanical properties of foams in the case of PC.

2.4.3 Variothermal Mold Temperature Control

The variotherm method involves tempering the tool to specifically influence certain properties of the injection molded component. This method is not limited to thermoplastic foam and is also used for conventional injection molding. Warmer tool surfaces provide many

Figure 2.28 Foamed molded part by Wittmann-Battenfeld made with (left) and without (right) a variotherm tool.

benefits for component properties, like increased mechanical values due to higher crystallinity, the reduction of internal stresses, and improved surface quality ranging to so-called *piano black surfaces* [4]. As Figure 2.28 shows, improved surface qualities are possible with the variotherm technique.

The method enables quick heating or cooling of the tool within an injection molding cycle. The advantage is that the polymer melt is injected into a heated tool ($T_{tool} > T_G/T_K$). This does not prevent the expansion of the blowing agent during the injection process in foam injection molding. However, it takes longer for the melt to freeze, so it can flow to the cavity wall before a freezing of the streaking patterns can occur. After the injection molding process, the tool is rapidly cooled to fixate the form. It is important that this be done before the gas diffuses out of the polymer melt and accumulates between the tool wall and component surface (i.e., the hammer effect). This effect can be counteracted by reducing the blowing agent pressure in the tool. The variotherm method can be employed on standard and on decompression tools [2].

Employing the variotherm foam injection molding method for thermoplastics leads to a reduction of surface roughness. Thus, despite longer cycle times, an economic gain is still achieved because the surface does not require postprocessing [4].

2.4.4 Sandwich Technique

The sandwich technique is an injection molding method that also can be used for a melt that does not contain a blowing agent. It is explained in more detail in Chapter 1. Generally, this technique is used when varying plastic components are used for core and skin of a component. For example, recycled plastics are incorporated in the core and do not have negative effects on the surface of the component.

In thermoplastic foam injection molding, the sandwich technique is used to create a compact outer layer and a foamed core. In addition to the already known advantages of foamed components, the improved surface quality is the main aspect. Because the melt containing blowing agent does not come into contact with the tool wall at any time during the process, and, thus, the surface roughness typical for thermoplastic foam injection molding does not occur, the components made in this manner display an improved surface quality. Essentially, the distribution between the core and the skin layer is difficult to achieve in sandwich components because it largely depends on the viscosity of the components and the flow path in the mold. In turn, the component viscosity depends upon the temperatures of the melt and tool [87]. Thus far, the precise correlations have not been examined because the process is strongly influenced by reciprocal effects. In principle, the precharge quantity is significant because it determines the remaining volume in the cavity. The shape of the core depends on the viscosity, tool temperature, and delay time until the injection of the core. Wide cores are achieved with high tool temperatures, which slow the freezing of the skin layer. Subsequently following melt can displace areas situated close to the wall. In addition to the core width, the skin layer directly correlates with the length of the core because both dimensions depend upon the volume of the core material. A qualitative model of varying core geometries for identical volume ratios of core and skin materials is shown in Figure 2.29.

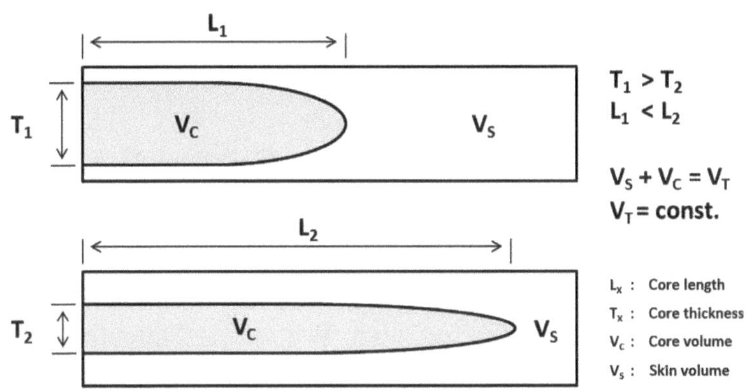

Figure 2.29 Illustration of varying core geometries for identical volume ratios of the core and skin materials in sandwich injection molding.

2.4.5 Physical Foaming Techniques

In general, physical blowing agents are mixed in during the plasticization process. In exceptional cases, the fluid is added during the injection phase (Optifoam®), or in a separate container (Ergocell). If a single-phase polymer-blowing agent mixture is available, the manner of incorporating a blowing agent plays only a minor role. The amount of blowing agent that can be added and the quality of homogenization differ significantly depending on the technique used.

2.4.5.1 *MuCell®*

The most widely used concept of physical foaming in injection molding is MuCell®, by Trexel Inc. (based in Woburn, Massachusetts). It was developed and patented [51,88–91] in the early 1990s at the Massachusetts Institute of Technology (MIT) in Cambridge, Massachusetts, and later was purchased by Trexel Inc. [48,92].

Nitrogen and CO_2 are mainly used as blowing agents, and are transferred into an overcritical state before addition [16,93]. The blowing agent is added via a nozzle in the plasticization cylinder during plasticization. Depending on the design, one or two injectors are designated for this, and are cascaded in correlation with the position of the screw [94]. A special screw geometry is used to achieve better mixing of the blowing agent fluid. It has an additional gas-mixing step that follows the three-zone area, and is meant to guarantee homogenous, single-phase mixing. Depending on the polymer employed, an additional nonreturn valve is installed between the plasticization and mixing zone to minimizing the loss of gas [2,14] (Figure 2.30).

2.4.5.2 *CellMould®*

In a process technological sense, the CellMould® technique by Wittmann Battenfeld, a firm based in Kottingbrunn, Austria, is similar to the previously described MuCell® technique. Verifiably, Wittmann Battenfeld already examined and developed this method of mixing blowing agents in the 1970s. The CellMould® technique is exclusively marketed in combination with Wittmann–Battenfeld injection molding machines, and it differs with regard to details (e.g., no second nonreturn valve) from machines available with MuCell® equipment,

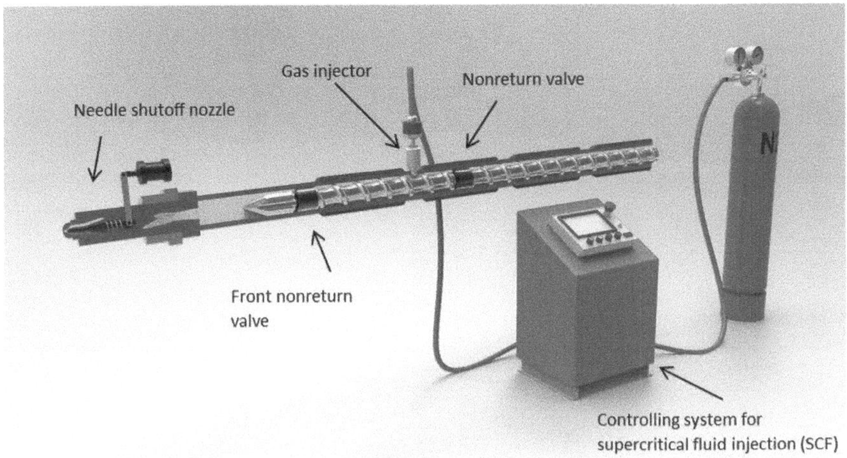

Figure 2.30 Schematic illustration of the MuCell® technique for foaming thermoplastics.

like those made by Engel (Schwertberg, Austria) and Krauss–Maffei (Munich, Germany) [95,96].

2.4.5.3 Ergocell

In 2001, Demag, based in Wetter, Germany, developed and introduced a new physical foaming technology for injection molding. In contrast to the addition of gas in the screw area, the blowing agent is injected via an additional module. This module is installed between the injection nozzle and plasticization cylinder. It consists of an introduction zone for the blowing agent fluid, a mixing part for the homogenization, and a connected shot-pot, in which the homogenized polymer-blowing agent mixture is stored under pressure until injection begins [97]. One advantage in comparison to the MuCell® technique is the independent piston pump speed, which is not linked to the movement of the screw. Thus, varying blowing agent concentrations and differing degrees of foaming can be achieved within the component. This method is especially suitable for polymers or compounds that are hard to foam because the mixing with gas can be decoupled from the plasticization process. Due to the high complexity in plant construction, retrofitting machines is not economic. The process-technological freedoms that come with this method are only obtained by great equipment investment [98,99]. The Ergocell technique is currently not provided by Demag (Figure 2.31).

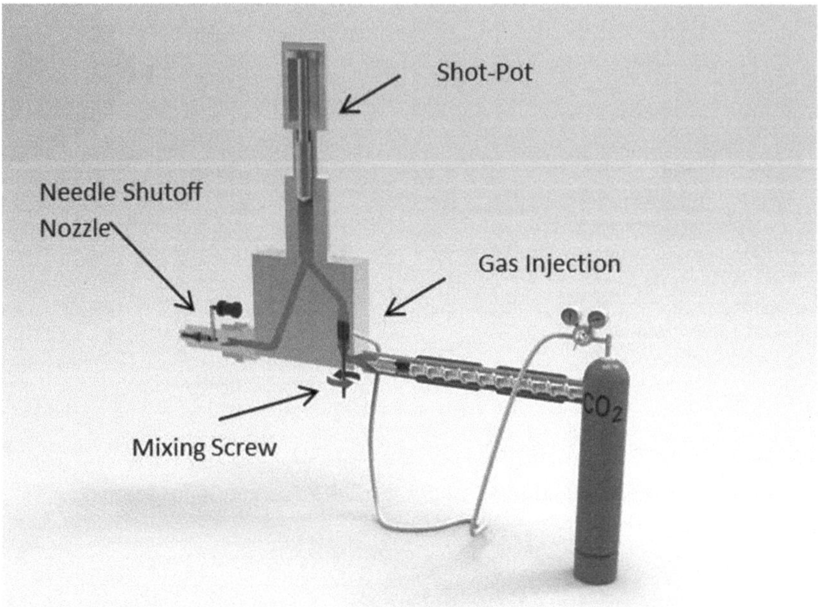

Figure 2.31 Schematic illustration of the Ergocell method for foaming thermoplastics.

2.4.5.4 *Optifoam®*

The Optifoam® method follows a similar approach as the Ergocell method. It was developed at the Institute of Plastics Processing (IKV) in Aachen [50,93,100]. Later, it was purchased by Sulzer Chemtech (Winterthur, Switzerland), and outsourced with the Static Mixers business division into the independent company Promix Solutions (Winterthur, Switzerland) in 2012 [101].

In contrast to the previously shown methods, this solution uses a different approach, and it was designed as a retrofitting system. An injection and mixing unit is installed between the needle shutoff nozzle and the plasticization cylinder. The injection and mixing unit consists of a centered torpedo, which causes the melt to pass through a narrow, annular gap. The surface shell of the torpedo and the inner cylinder are made of a sintered metal ring, through which the blowing agent is injected into the melt. Subsequently, the polymer-gas mixture is homogenized in a static mixer and then transferred into a single-phase solution. In the Optifoam® method, the addition of the blowing agent takes place during the injection process [48,102,103]. At present, this technology is no longer available for injection molding machines (Figure 2.32).

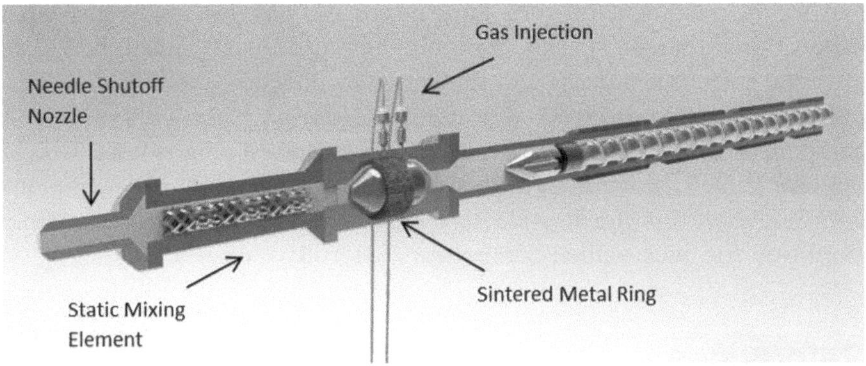

Figure 2.32 Schematic illustration of the Optifoam® method for foaming thermoplastics.

2.4.5.5 *ProFoam*

Since 2008, the Institute of Plastics Processing in Aachen has been developing another method to foam polymers directly in injection molding. While the other methods inject the blowing agent into the already melted polymer, plastic granulate subject to pressure and heat is preloaded in the ProFoam method, similar to a batch process, and is subsequently plasticized. Like the Optifoam® system, it is a retrofitting solution, and it is primarily suitable for foaming thermally sensitive polymers, which possess a high gas diffusion rate. The ProFoam system is installed on the hopper of the injection molding machine. It has a pressure chamber that is sealed with two locks. By means of diffusion, gas is incorporated into the polymer granulate in this volume, and then held under pressure until the injection process initiates. There must be a seal on the side of the motor of the plasticization screw. In comparison to the direct gassing method, the maximum amount of blowing agent is limited based on the low solubility of the gas in the solid granulate [104–106].

2.4.5.6 *SmartFoam®*

The SmartFoam® method was developed by Stieler Kunststoff Service GmbH (Goslar, Germany), and is not directly integrated into the injection molding machine. A hot runner system with needle shutoff nozzles is needed because the blowing agent is injected into the tool. This method is very similar to the Optifoam® system because the loading of the gas takes place during the injection process [107]. In this case, a static mixer is left out, leading to a poorly homogenized mixture and the development of very large pores. Partial gas loading is what makes this technique special, for a blowing agent is not added before or after the injection process. As in sandwich injection molding, the gas-loaded polymer is injected as a core into the polymer without gas. A component with a compact skin layer and a foamed core results. Due to the problematic blowing agent-polymer homogenization, this technique is not suitable for the manufacture of microfoams. Thus, it is mainly employed for thick-walled components [81,108].

References

[1] Xu X, Park CB. Injection foam molding. In: Isayev A, Kamal MR, editors. Injection molding. München: Hanser-Verlag; 2009.
[2] Altstädt V, Mantey A. Schaumspritzgießen. München: Hanser-Verlag; 2010.
[3] Johannaber F, Michaeli W. Handbuch Spritzgiessen. München: Hanser-Verlag; 2004.
[4] Spörrer A. Leichte Integralschäume durch Schaumspritzgießen mit optimierten Werkstoffen und variothermen Werkzeugen//Evaluierung des Schäumverhaltens eines mittels Kern-Schale-Teilchen zähmodifizierten, rezyklierbaren amorphen PolymersDissertation. Bayreuth: TuTech-Verlag; 2010.
[5] Bürkle E, Wobbe H. Die bessere Alternative zum Kompaktspritzgießen? Kunststoffe 2014;2:44–6.
[6] Steinbichler G. Gehobene Softtouch-Qualität in einem Arbeitsgang. Kunststoffe 2006;7:76–7.
[7] Kühn-Gajdzik J. Amorphe und teilkristalline Mikroschäume im SpritzgießverfahrenDissertation. Kassel: Kassel University Press; 2011.
[8] Stange J. Einfluss rheologischer Eigenschaften auf das Schäumverhalten von Polypropylenen unterschiedlicher molekularer StrukturDissertation. Erlangen: Jens Stange; 2006.
[9] Colton JS, Suh NP. The nucleatuion of microcellular thermoplastic foam with additives: part I: theoretical considerations. Polym Eng Sci 1987; 7:485–92.
[10] Arefmanesh A, Advani SG, Michaelides EE. A numerical study of bubble growth during low pressure structural foam molding process. Polym Eng Sci 1990;20:1330–7.
[11] Eaves D. Handbook of polymer foams. Shawbury, UK: Smithers Rapra Press; 2004.
[12] Hansen S. Schaumspritzgießen in der Praxis (Konferenz Thermoplastische Schäume): Aachen; 2003.

[13] Michaeli W, Haberstroh E, Lorenz J, Habibi Naini S, Cramer A. Kunststoffschäume (IKV-Kollequium): Aachen; 2004.
[14] Okamoto KT. Microcellular processing. München, Cincinnati: Hanser-Verlag; 2003.
[15] Bledzki AK, Kirschling H, Steinbichler G, Egger P. Microcellular moulding with gas counter pressure (21st Annual Meeting of the Polymer Processing Society), Leipzig; 2005.
[16] Michaeli W, Pfannschmidt O, Habibi-Naini S. Wege zum mikrozellularen Schaum – Grundlagen zu Material und Prozessführung. Kunststoffe 2002;6:48–51.
[17] Egger P, Fischer M, Kirschling H, Bledzki A. Serienfeste Vielseitigkeit beim MuCell-Spritzgießen – Ein Statusbericht für die Praxis Teil 1. Kunststoffe 2005;95(12):66–70.
[18] Egger P, Fischer M, Kirschling H, Bledzki A. Serienfeste Vielseitigkeit beim MuCell-Spritzgießen – Ein Statusbericht für die Praxis Teil 2. Kunststoffe 2006;96(1):72–6.
[19] Schröder U. Pole Position - Mit Mucell-Technik Spirtzgießen. Kunststoffe 2003;93(10):30–3.
[20] Schwarz O, Ebeling F-W, Furth B. Kunststoffverarbeitung, Vogel Verlag: Würzburg; 2002.
[21] Wahlen L. Luvopor – Chemisches Treibmittel im Spritzgießprozess (2. Workshop Polymere Mikroschäume), Kassel; 2005.
[22] Michaeli W, Opdenwinkel K. Schäumen ohne Chemie ist keine Magie. Plastveratbeiter 2009;60(1):24–5.
[23] Lübke G, Holzberg T. Dünnwandige Bauteile – Neue Treibmittel-Entwicklungen im Bereich Spritzgießen (IKV-Seminar zur Kunststoffverarbeitung): Aachen; 2003.
[24] Kirschling HMM. Mikroschäume aus Polycarbonat, Herstellung - Struktur - EigenschaftenDissertation, PPH ZAPOL Dmochowski, Sobczyk Spółka Jawna: Kassel; 2009.
[25] Peng J, Turng L-S, Peng X-F. A new microcellular injection molding process for polycarbonate using water as the physical blowing agent. Polym Eng Sci 2012;52(7):1464–73.
[26] Sorrentino L, Aurilia M, Iannace S. Polymeric foams from high performance thermoplastics. Adv Polym Technol 2011;30(3):234–43.
[27] Park C, Behravesh A, Venter R. Low density microcellular foam processing in extrusion using CO_2. Polym Eng Sci 1998;38(11):1812–23.
[28] Saiz-Arroyo C, de Saja J, Rodriguez-Perez M. Production and characterization of crosslinked low-density polyethylene foams using waste of foams with the same composition. Polym Eng Sci 2011;52(4):751–9.
[29] Zhai W, Wang H, Yu J, Dong JH. Cell coalescence suppressed by crosslinking structure in polypropylene microcellular foaming. Polym Eng Sci 2008;48(7):1312–21.

[30] Rohleder M. Dynamisches Lastverhalten von mikrozellularen Polycarbonatschäumen, Morphologie, Energieabsorption und Temperatureinfluss-Dissertation. Kassel: Kassel University Press; 2015.
[31] Bledzki A, Faruk O, Kirschling H, Kühn J, Jaszkiewicz A. Microcellular polymer and composites part II: properties of different types of microcellular materials. Polymery 2007;52(1):3–12.
[32] Park CB, Zheng W. Clay exfoliation and content effect on nylon nanocomposite foams (63th Annual Technical Conference): Boston; 2005.
[33] Kumar V, Suh N. A process for making microcellular thermoplastic parts. Polym Eng Sci 1990;30(20):1323–9.
[34] Klemperer D, Sendijarevic V. Polymeric foams and foam technology. München: Hanser-Verlag; 2004.
[35] Gibson LJ, Ashby MF. Cellular solids. Camebridge: Camebridge University Press; 1997.
[36] Hilyard NC. Mechanics of cellular plastics. London: Applied Science Publishers; 1982.
[37] Nimmer R, Stokes V, Ysseldyke D. Mechanical properties of rigid thermoplastic foams - part II: stiffness and strength data for modelling polypropylene oxide foams. Polym Eng Sci 1988;28(22):1501–8.
[38] Throne JL. Thermoplastic foams. Hinkley: Sherwood Publishers; 1996.
[39] Müller N. Spritzgegossene Integralschaumstrukturen mit ausgeprägter Dichtereduktion (Dissertation): Erlangen; 2006.
[40] Xu J, Kishbaugh L. Simple modeling of the mechanical properties with part weight reduction for microcellular foam plastics. J Cell Plastics 2003;39(1):29–47.
[41] Semerdjiev S. New structural foam molding process offers improved surface appearance. J Cell Plastics 1976;12(3–4):118–23.
[42] Popov N, Semerdjiev S. Technologische Möglichkeiten des Strukturschaum-Spritzgießens mit Gasgegendruck, Fachbeitrag 1.Teil. Kunststoffberater 1980;25(7–8):23–5.
[43] Popov N, Semerdjiev S. Technologische Möglichkeiten des Strukturschaum-Spritzgießens mit Gasgegendruck, Fachbeitrag 2.Teil. Kunststoffberater 1980;25(9):54–6.
[44] Semerdjiev S, Piperov N, Popov NM. Das Gasgegendruck-Gießverfahren zum Herstellen von thermoplastischen Strukturschaumteilen. Kunststoffe 1974;64(1):13–15.
[45] Michaeli W, Cramer A. Bessere Oberflächen beim Schaumspritzgießen. Kunststoffe 2009;96(12):21–7.
[46] Poetsch G, Michaeli W. Injection moulding. München: Hanser-Verlag; 2007.
[47] Berry M. Microcellular injection molding. In: Kutz M, editor. Applied plastics engineering handbook. Burlington: Elsevier Science; 2011.
[48] Habibi-Naini S. Keine Scheu vorm Schäumen. Kunststoffe 2004;94 (9):191–4.

[49] Jaroschek C. Spritzgießen für Praktiker. München: Hanser-Verlag; 2013.
[50] Michaeli W, Habibi-Naini S. Grundlagen des Schäumens mit physikalischen Treibfluiden (SKZ-Tagung: Physikalisches Schäumen von Thermoplasten beim Spritzgießen): Würzburg; 2002.
[51] Park C, Suh N, Baldwin D. Method for providing continuous processing of microcellular and supermicrocellular foamed materials US5866053.
[52] Fleischli M, Streiff F. Neue Erkenntnisse zum Mischen und Dispergieren von Flüssigkeiten mit großen Viskositätsunterschieden in statischen Mischern. Chem-Ing-Tech 1990;62(8):650−4.
[53] Müller W. Mechanische Verfahrenstechnik und ihre Gesetzmäßigkeiten, Mechanische Grundoperationen und ihre Gesetzmäßigkeiten. Berlin: De Gruyter-Verlag; 2014.
[54] Heusser R. Operating out of sight. Kunststoffe Int 2013;10:129−31.
[55] Pahl MH, Muschelknautz E. Einsatz und Auslegung statischer Mischer. Chem-Ing-Tech 1979;51(5):347−64.
[56] Rohleder M. Analyse des Deformations- und Bruchverhaltens von Polycarbonat mit mikrozellularer Hohlraumstruktur (Unveröffentlichte Diplomarbeit): Kassel; 2007.
[57] Rohleder M, Bledzki AK, Kirschling H. Correlation between injection moulding parameters, morphology and properties of microcellular polycarbonate (Blowing Agent and Foaming Processes 2009): Hamburg; 2009.
[58] Almstedt S. Morphologie und Bruchverhalten von Mikroschäumen aus Polycarbonat, hergestellt mit Gasgegendruckverfahren und Präzisionsöffnen (Unveröffentlichte Diplomarbeit): Kassel; 2006.
[59] Braun S. Gezielte Optimierung des Eigenschaftsniveaus von Polycarbonat durch Modellierung, Herstellung und Untersuchung mikrozellularer Schaumstrukturproben (Unveröffntlichte Diplomarbeit): Kassel; 2004.
[60] Bledzki AK, Rohleder M, Kirschling H. Influence of different blowing agents and injection moulding processing parameters on the properties of microcellular polycarbonate (Blowing Agent and Foaming Processes 2009): Hamburg; 2009.
[61] Wahlen L. Properties and applications of chemical blowing agents (Blowing Agent and Foaming Processes 2007): Frankfurt; 2007.
[62] Reedy ME. Improved nucleation of polystyrene foams (Foams 2008 - TopCon), Charlotte; 2008.
[63] Barth C. Modifikation von Polycarbonat durch gezielte Beeinflussung der Morphologie (Dissertation): Düsseldorf; 1999.
[64] Potente H, Moritzer E, Obermann C. Foam formation in gas-assisted injection molded parts: theoretical and experimental considerations. Polym Eng Sci 1996;36(16):2163−71.
[65] Crank J, Park GS. Diffusion in polymers. London and New York: Academic Press; 1968.
[66] Vieth WR. Diffusion in and through polymers. München: Hanser-Verlag; 1991.

[67] Tang M, Du T-B, Chen Y-P. Sorption and diffusion of supercritical carbon dioxide in polycarbonate. J Supercritical Fluids 2004;28 (2–3):207–18.

[68] van Krevelen DW. Properties of polymers. Amsterdam: Elsevier; 1976.

[69] Kohlshorn U, Traut H. Prozess- und Qualitätsüberwachung bei der Produktion mikrozellular geschäumter Bauteile (SKZ-Schaumseminar): Würzburg; 2006.

[70] Cramer A. Analyse und Optimierung der Bauteileigenschaften beim Thermoplast-Schaumspritzgießen, Analysis and optimisation of the part properties in thermoplastic foam injection moulding Dissertation. Aachen: Verlag-Mainz; 2008.

[71] Bledzki AK, Kirschling H, Steinbichler G, Egger P. Microcellular moulding with gas counter pressure using physical blowing agent (Cellular Metals and Polymers): Fürth; 2004.

[72] Florez Sastre L. Analysis of the mechanical response of rigid thermoplastic foams (Blowing Agent and Foaming Processes 2007): Frankfurt; 2007.

[73] Spörrer A, Altstädt V. Controlling morphology of injection molded structural foams by mold design and processing parameters. J Cell Plastics 2007;43(4–5):313–30.

[74] Rodrigue D, Leduc S, Tovar-Cisneros C. The effect of mould temperature on the morphology of injection moulded HDPE structural foams (Blowing Agent and Foaming Processes 2007): Frankfurt; 2007.

[75] Reza Barzegari M, Rodrigue D. The effect of processing conditions on the morphology of injection moulded structural foams (Blowing Agent and Foaming Processes 2008): Berlin; 2008.

[76] Florez Sastre L. Einfluss der Schaummorphologie auf die mechanischen Eigenschaften von Kunststoffstrukturschäumen Dissertation, Mainz, Aachen; 2011.

[77] Eckardt H. Smooth foam and smooth surface - a contradiction? (Blowing Agent and Foaming Processes 2009): Hamburg; 2009.

[78] Eckardt H. Schaumspritzgießverfahren - Theorie & Praxis. Kunststoffberater 1983;28(1–2):26–32.

[79] Menges G, Michaeli W, Mohren P. Spritzgießwerkzeuge Auslegung, Bau und Anwendung, Hanser, München: Wien; 2007.

[80] Mantey A, Sandler JKW, Altstädt V, Kunz D. Foam injection-moulding of polymer components with a significant density reduction (21st Annual Meeting of the Polymer Processing Society): Leipzig; 2005.

[81] Stieler U. Schäumen im Werkzeug. Kunststoffe 2008;98(9):68–71.

[82] Heim H-P, Schnieders J, Jarka S, Gövert S. Geschäumte Rippen für dünnwandige Teile. Kunststoffe 2011;101(11):48–50.

[83] Heim H-P, Tromm M. Formation of morphology as a function of process control by foam injection molding of a functionally graded component (ANTEC 2014), Las Vegas; 2014.

[84] Heim H-P, Tromm M. Injection molded components with functionally graded foam structures - Procedure and essential results. J Cell Plastics 2015;1–21.
[85] Bledzki A, Kirschling H, Steinbichler G, Egger P. Polycarbonate microfoams with a smooth surface and higher notched impact strength. J Cell Plastics 2004;40(6):489–96.
[86] Hellerich W, Harsch G, Haenle S. Werkstoffführer Kunststoffe: Eigenschaften, Prüfungen, Kennwerte. München: Hanser-Verlag; 2001.
[87] Moritzer E. Beschreibung und Vorhersage der Kern-/Hautverteilung bei Sandwichbauteilen mit Hilfe der Ähnlichkeitstheorie, Schlussbericht: Paderborn; 2011.
[88] Park C, Suh N, Baldwin D. Method for providing continuous processing of microcellular and supermicrocellular foamed materials US6051174.
[89] Suh N, Hermann T. Gear throttle as a nucleation device in a continuous microcellular extrusion system US6005013.
[90] Pierick DE, Anderson JR, Cha SW, Stevenson JF, Laing DE. Injection molding of microcellular material EP1264672.
[91] Baldwin D, Suh N, Park C, Cha S. Supermicrocellular foamed materials USRE37932.
[92] Fischer M. Foammelt, MuCell - ein neuer Impuls in der Strukturschaumtechnologie, Schwertberg.
[93] Michaeli W, Pfannschmidt O, Habibi-Naini S. Spritzgiessen mikrozellulärer Schäume. Kunststoffe 2002;92(8):56ff.
[94] Becker O, Paruchuri S. Schaumspritzgießen mit physikalischen Treibmitteln (MuCell) (Konferenz Thermoplastische Schäume): Aachen; 2001.
[95] Doriat C. Wiederauferstehung eines Traditionsunternehmens. Kunststoffe 2008;98(8):55–7.
[96] Stein A. Zusammen wachsen. Kunststoffe 2009;99(11):20–1.
[97] Sauthof R. Potenziale des physikalischen Schäumens mit ErgoCell (Seminar Kunststoffschäume): Aachen; 2006.
[98] Buchmann M, Busch S, Jäger A, Klenz R. Nicht die Blasengröße ist entscheidend. Kunststoffe 2005;95(10):216–20.
[99] Pahlke S. Schaumspritzgießen mit physikalischen Treibmitteln – Maschinentechnik, Nutzen und Grenzen (Seminar Thermoplastische Schaumstoffe Verarbeitungstechnik und Prozessanalyse): Aachen; 2003.
[100] Habibi-Naini S, Schlummer C, Lecher J, Pany M. Schaum-Schlösser. Kunststoffe 2006;96(3):128–33.
[101] NN. Sulzer Kunststoffverarbeitung wird Promix Solutions. Kunststoffe 2012;102(5).
[102] Habibi-Naini S, Schlummer C. Geschäumtes Flüssigsilikon eröffnet neue Wege. Kunststoffe 2006;96(4):91–3.
[103] Bothur C. Hohes Potenzial. Plastveratbeiter 2004;55(10):33–6.

[104] Michaeli W, Hopmann CH, Obeloer D. Examinations on the influencing factors on the foamability using the ProFoam process (ANTEC 2011 – Proceedings of the 69th Annual Technical Conference & Exhibition): Boston; 2011.
[105] Michaeli W, Neuß A, Obeloer D. ProFoam – Einfach, robust, flexibel Schäumen (VDI-Jahrestagung Spritzgießen 2011), Baden-Baden; 2011.
[106] Michaeli W, Obeloer D. A new process for the injection moulding of foamed parts with physical blowing agents (Blowing Agents and Foaming Processes): Hamburg; 2009.
[107] Stieler U. Smart Foam: a new physical foaming process for injection moulding (Blowing Agent and Foaming Processes 2008): Berlin; 2008.
[108] Stieler U. Mehr als nur Hohlräume. Kunststoffe 2011;101(11):48–50.

3 Gas- and Fluid-Injection Technique

Hans-Peter Heim
Universität Kassel, Institut für Werkstofftechnik, Kunststofftechnik, Kassel, Germany

3.1 Introduction

Fluid injection technology encompasses all methods in which part of the melt is displaced directly after the injection of the melt into the injection molding tool by injecting a fluid (i.e., water) or gas (i.e., N_2), therefore creating a hollow space in the component. In doing so, fluid injection technology offers, among others, the following technical benefits:

- There is greater design freedom in terms of the design of the molded part and the tool.
- Sink marks and warpage can be reduced by means of an according position of the hollow space.
- Owing to the formation of hollow profiles with a large cross section, better stiffness values can be achieved (area moment of inertia).
- In contrast to compact molded parts of the same volume, a reduction of the cycle time and material and weight reductions are possible.
- In the case of molded parts with a large surface area, a reduction of the clamp force is possible.
- Long flow paths can be realized very well.

The beginning of fluid technology can be traced to a method published in 1943 by John R. Hobson [1]. This method consists of the injection of gas into a liquid melt. Hobson's invention is very similar to the so-called core-pull method of gas injection technology (GIT), a variant of GIT that was first applied in practice on a larger scale several decades later. The patents developed by Friedrich and Morbach in the 1970s are mentioned in various publications as a role model for today's

Specialized Injection Molding Techniques.
DOI: http://dx.doi.org/10.1016/B978-0-323-34100-4.00003-1

variants of GIT [2,3]. After being only gradually applied in practice in the beginning, GIT gained significance at the end of the 1980s, and especially in the 1990s. At the end of the 1990s, water injection technology (WIT) was additionally developed and extensively analyzed in numerous dissertations.

Due to the advancing of this technology in the 1990s, an increasing number of products were introduced into the market, and plastics processers gained abundant experience in terms of tool design, process control, and material selection. In the application-related departments of raw material and machine manufacturers, as well as at many universities, research and development projects were carried out with the aim of understanding GIT and providing methods for tool and process design. Retrospectively, it can be said that the time period from the 1990s to the beginning of the twenty-first century was the most dynamic period for fluid injection technology in terms of new technological developments and publications, as well as with regard to research and development. The number of dissertations that were created in this time alone is remarkable.[1]

In the meantime, GIT and WIT have been included in the numerous available injection molding methods under various manufacturer-specific names. These include the following:

- GIT (Bauer Comp Holding GmbH, PME fluidtec GmbH; Munich, Germany)
- Airmould/Aquamould (Wittmann Battenfeld GmbH & Co. KG; Torrington, CT/Meinerzhagen, Germany)
- Cinpres (Cinpres Gas Injection Ltd.; Winsford, Cheshire, UK)
- Gasmelt/Watermelt (Engel Austria GmbH; Schwertberg, Austria)
- GID/WID (Krauss Maffei AG; Munich, Germany)
- Fluid Injection Technique, FIT (Arburg GmbH & Co. KG; Lossburg, Germany)
- Gas Assisted Moulding, GAM (Rutland Plastics Ltd.; Oakham, Rutland, UK)
- Assisted Injection Moulding, AIM (Maximator GmbH; Nordhausen, Germany)

---

[1]See dissertations by Lanvers AP (1993), Hansen M (1996), Rennefeld C (1996), Findeisen H (1997), Schröder T (1997), Moritzer E (1997), Heim H-P (2001), Hopmann C (2001), Kleba I (2001), Wehr H (2002), Jüntgen T (2004), Ridder H (2005), Pretel G (2006), Lettowsky C (2009).

For many years, the simulation of mold-filling behavior, cooling, the resulting sizes and orientations in the component, and warpage and shrinkage have been inherent parts of the development process of injection molding tools and injection molded parts. Customarily, finite element methods (FEMs), or finite volume methods (FVMs) are employed with commercially available software (that is, Moldflow®, Moldex®, SIGMASOFT®, Cadmould®, or Solidworks®[2]). The calculation of the GIT process was only conditionally reliable at the beginning of the 21st century, meaning the calculations of correct hollow spaces and the resulting cooling conditions, shrinkage, and warpage were not precise enough to predict the behavior of complex components. Meanwhile, FVM-based software enables a very good simulation of fluid injection technology and the resulting component properties.

Nonetheless, a simulation software can deliver a useful result only if the designer provides both a good component and good tool design, and if the calculation results can be interpreted purposefully. For this, sound knowledge of the injection molding method is required.

In this chapter, GIT is dealt with in detail. In order to provide clearer illustrations, the focus will be on the partial filling method. The partial filling method of GIT encompasses the essential questions that are necessary for the comprehension of fluid injection technology. In the last section of this chapter, the different method variants and their special features are explained. Examples include the overflow cavity method of GIT, WIT, and the GITBlow process. The sections of this chapter have been structured as follows:

- Explanation of the GIT process and illustration of typical gas-guiding geometries
- Sample illustrations and short explanations of typical GIT molded parts
- Continuing discussions to achieve in-depth insight into the formation of the hollow space or gas bubble.
- Description of the influences of material and process parameters.
- Illustrations of recommendations for the design of GIT components
- Illustration of variants of fluid injection technology and combinations of methods.

---

[2]See product information by Autodesk Inc., California; CoreTech System Co., Ltd., Taiwan; SIGMA Engineering GmbH, Germany; Simcon Software GmbH, Germany; Dassault Systèmes SOLIDWORKS Corp., France.

3.2 Description of the GIT Procedure

In the GIT partial filling method, only a portion of the tool cavity is filled with melt. Gas is subsequently injected. Thus, the holding pressure conventional for injection molding is not used in this method, and the machine control switches to gas injection directly after melt injection (in some case, gas injection can be started shortly before the completion of the injection phase if the machine control enables it). The injection of gas can take place either via a so-called machine nozzle or via an injection tool (see Figure 3.1) [4]. Both methods have specific advantages and disadvantages [5]. Today, injection tools are common. Because the melt is displaced from the fluid core of the molded part

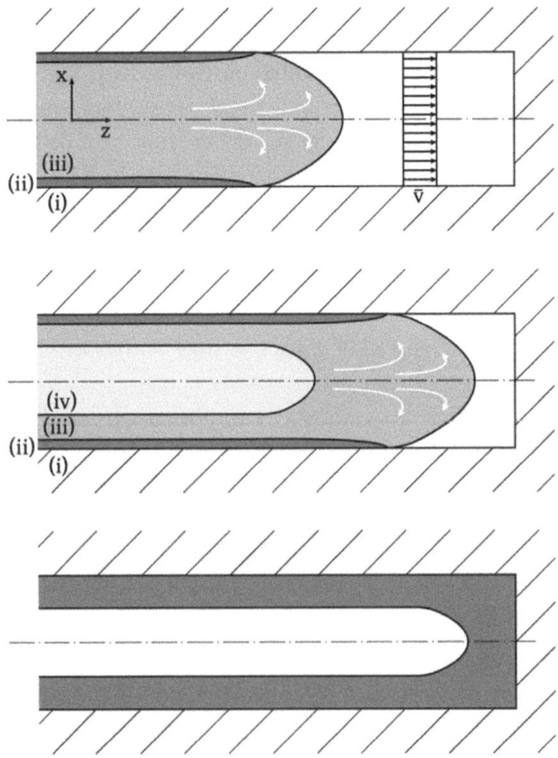

Figure 3.1 Procedure of the GIT partial filling method. First step: partial filling with melt; second step: gas is injected and melt is moved; third step: filling of the mold is completed. (i) Mold wall, (ii) frozen melt layer, (iii) melt, (iv) gas bubble, $\bar{v}$, average filling velocity.

and is replaced by gas when the hollow space forms, a volume that equals the size of the displaced material must be available in the molded part [6].

In contrast, a complete, volumetric filling of the mold is carried out in the overflow cavity method. The moment that gas injection begins, an attached volume is opened in the tool. During injection, the melt is displaced into this area via the guiding geometry [5]. In the overflow cavity method, it is possible (but not mandatory) to apply a holding pressure. This procedure is especially favored for complex component geometries or inconvenient gate locations. Further explanation of this topic can be found in the final section of this chapter.

After the formation of the hollow space, the gas-holding phase follows during the cooling phase. A sustained gas pressure is used to achieve a compensation for shrinkage. In order to exemplify the chronological cycle of the process and the definition of additional designs, the GIT injection molding cycle is illustrated in Figure 3.2.

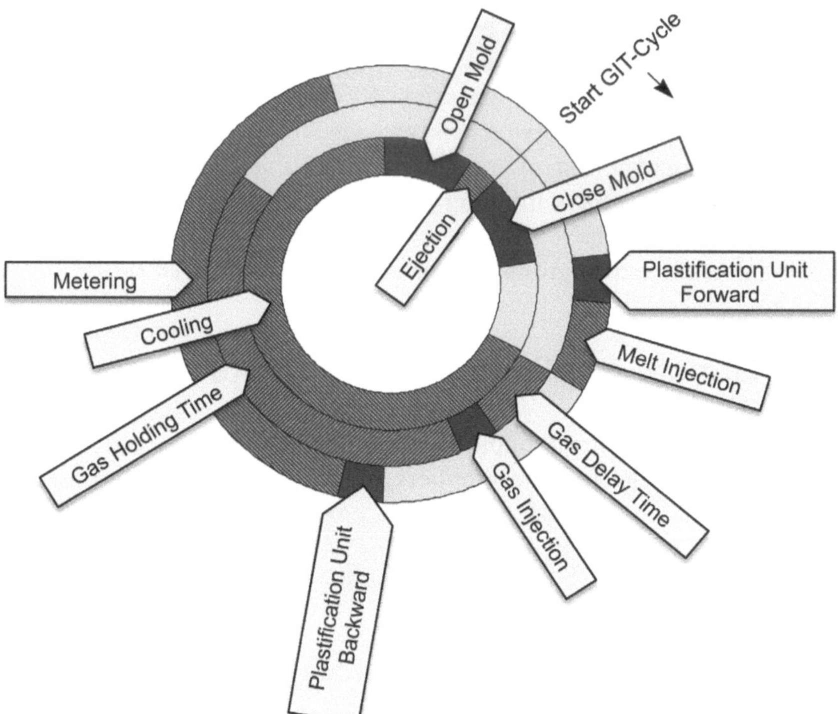

Figure 3.2 Chronological listing of the GIT injection molding cycle.

It has become common to differentiate between the following three geometric groups when employing GIT [6]:

- *Bar-shaped, thick-walled molded parts.* Molded parts with large wall thicknesses H with a relatively low component width B (reference values: $H > 4$ mm, $B/H < 3$)
- *Thin-walled molded parts with gas-guiding ribbing.* An alternative to conventional ribbing in plastic parts; design and haptic are more favorable than the application of conventional ribbing
- *Thin-walled molded parts with partially thick areas.* Molded parts with mostly circumferential thick-walled areas, either for the improvement of the stiffness of the component or due to the design of the component

The three geometric groups named here display areas with more or fewer material accumulations that are either unsuitable or only conditionally suitable for manufacturing using compact injection molding. By gas injection molding, these geometric groups can be manufactured efficiently because material accumulations are avoided in the thick-walled areas due to the hollow space. Thus, the cooling times and sink marks can be significantly reduced. Moreover, the pressure on the inside of the component, which is provided by GIT, can be sustained over a very long period of time. It replaces the conventional holding pressure in the injection molding cycle and leads to components with low levels of warpage.

The three geometric groups are presented in Table 3.1.

3.3 Typical Applications and Tasks

GIT finds widespread application in the automotive and furniture industries, whose large-scale suppliers are small and medium-sized plastic processors. On the whole, GIT is employed in all sectors, as is shown in the examples in Table 3.2.

Table 3.1 Geometry Groups for GIT Molded Parts

	Bar-shaped, thick-walled molded parts	Thin-walled molded parts with gas-guiding ribbing	Thin-walled molded parts with partially thick spots
Typical geometry[a]	Car door handle 	Cover plate 	Car cover 
Representative cross sections[b]	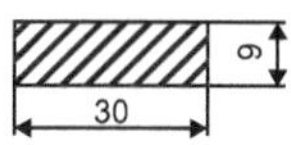	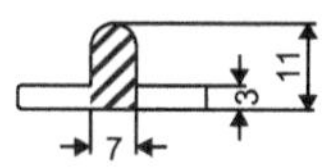 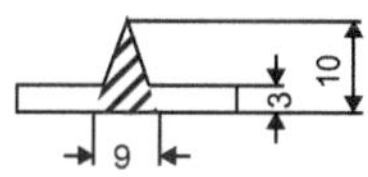	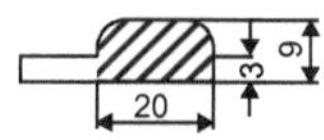
A_{rep}[c]	270 mm^2	70 mm^2	170 mm^2
Cross sections of the hollow space[d]	30%–40%	40%–50%	35%–45%

[a]Exemplary component.
[b]Simplified, schematic depiction of typical geometric cross sections; the shading marks the area of the cross section that is relevant for the formation of a hollow space A_{rep}.
[c]A_{rep} is the calculated area of the representative cross section and is exemplary for typical cross sections.
[d]Reference values for maximum achievable hollow spaces (in percent) for the relevant cross section area A_{rep}.

Table 3.2 Typical Examples of the Application of GIT

Medical technology: Infusion system	**Car: Door handle**
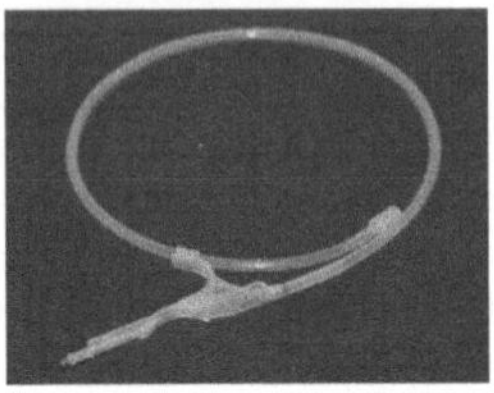 Reasons for employing GIT: – One-step process including tube – No joining of molded and extruded parts necessary	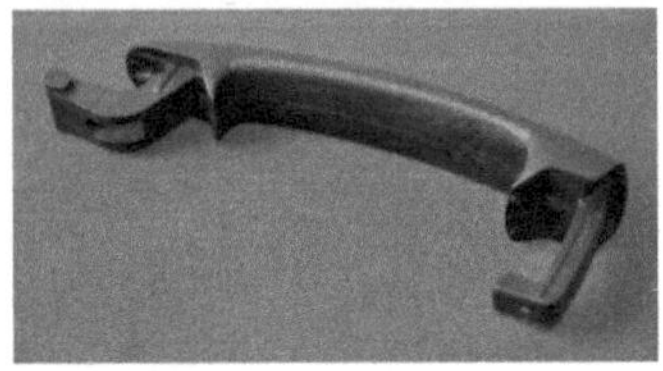 Reasons for employing GIT: – Thick-walled, molded part with hollow core – Less shrinkage, less cooling time, less raw material than is needed for solid molded part
Car: Lid of glove compartment	**Furniture: Armrest of office chair**
 Reasons for employing GIT: – Partially thick-walled for better haptic – Avoid warpage, increase stiffness	 Reasons for employing GIT: – Thick-walled part – High stiffness and better haptic than ribbings – Low cooling time compared to solid molded part
White goods: Casing part	**Car: Handle**
 Reasons for employing GIT: – Surrounding thick-walled area – Increase stiffness and avoid warpage	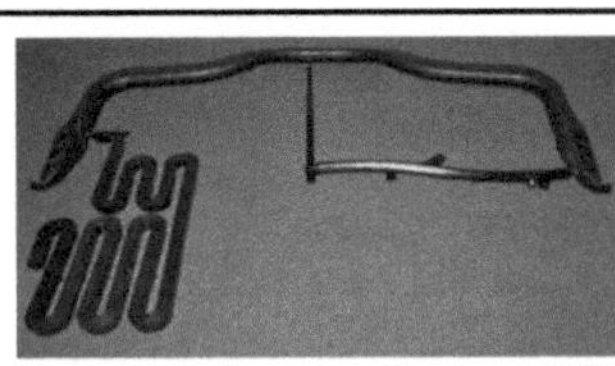 Reasons for employing GIT: – Thick-walled, molded part with hollow core – Less cooling time and less raw material than is needed for solid molded part

Car: Clutch pedal	Furniture: Bathtub handle
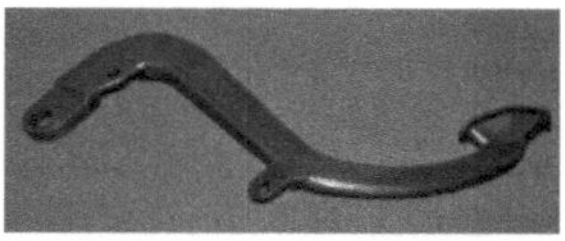	
Reasons for employing GIT: – Thick-walled, molded part with hollow core – Increase stiffness – Better design than with ribbings	Reasons for employing GIT: – Thick-walled, molded part with hollow core – Better haptic than with ribbings

3.4 Mechanisms of Gas Penetration

In the following examples, the influence of the setting parameters of the GIT injection molding process are explained. In later sections of this chapter, the focus shifts to the geometric and material parameters.

The GIT procedure can be divided into the following three process steps:

1. Injection/partial filling
2. Gas injection/remaining filling
3. Cooling/formation of the quality features of the component

The three steps are shown in Table 3.3, with the essential process step results and the most important process parameters of the associated processing step. It is important to note that the process step results of a previous process step are considered the initial values for the succeeding step.

3.4.1 Injection/Partial Filling

Principally, the partial filling is identical to that of filling compact components, meaning, as in standard injection molding, the fountain flow typical for highly viscose plastic melts, which stick to the walls, is available. A frozen marginal layer forms (see Figure 3.3). The according courses of the flow speed v, the shear speed $\dot{\gamma}$, and the viscosity η are shown in Figure 3.4.

Partially filling the cavity creates the initial condition for the formation of a hollow space using gas injection. It does this in two ways: first, by the amount of injected melt (filling degree); and second, by the associated melt temperature and temperature distribution.

Table 3.3 Process Steps and Their Process Parameters

Process step	Process step result	Influencing factors	
Cooling/ formation of the quality features of the component	Dimensions	Gas bubble length	l_{Gas}
	Surface properties, etc.	Gas bubble cross section	A_{Gas}
		Gas bubble volume	V_{Gas}
		Material temperature after injection	T_{mE}
		Mold wall temperature	T_{WZ}
		Second gas pressure step	p_{G2}
		Gas holding time	t_{GH}
Gas injection/ remaining filling	Gas bubble length	Injected mass	m_E
	Gas bubble cross section	Material temperature after injection	T_{mE}
	Gas bubble volume	Temperature distribution after injection	dT_{mE}
		Gas delay time	t_{VG}
		First gas pressure step	p_{G1}
		Mold wall temperature	T_{WZ}
Injection/partial filling	Injected mass	Material temperature in front of the screw	T_{mSV}
	Material temperature/ temperature distribution after injection	Temperature distribution in front of the screw	dT_{mSV}
		Dosed mass in front of the screw	m_{SV}
		Injection speed	v_E

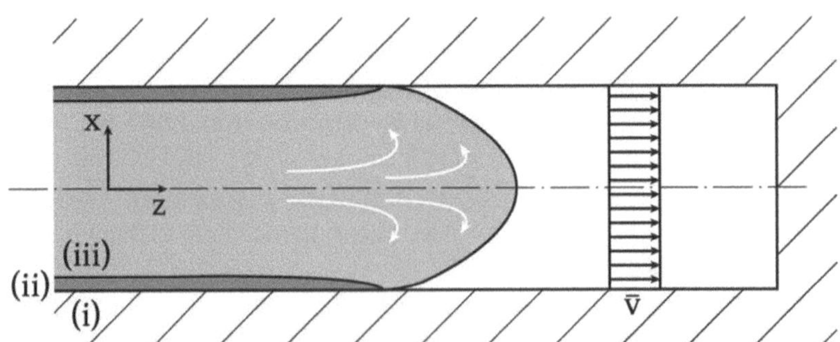

Figure 3.3 Fountain flow during melt injection: (i) mold wall, (ii) frozen melt layer, (iii) melt, $\bar{v}$, average filling velocity.

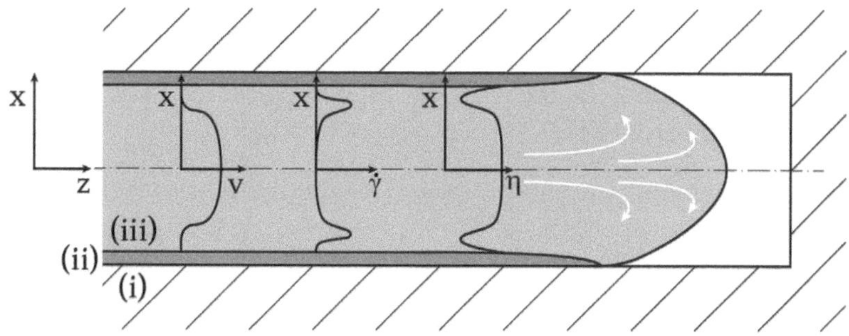

Figure 3.4 Courses of flow speed v, shear speed $\dot{\gamma}$, and viscosity η during partial filling of the cavity: (i) mold wall, (ii) frozen melt layer, (iii) melt.

3.4.2 Gas Injection/Remaining Filling

The hollow space created by gas injection is characterized by the gas canal cross section [or, alternatively, the remaining wall thickness (RWT) in the area of the created hollow space] and by the length of the gas canal. The quantitative influence of the process parameters on the canal cross section and length is assessed differently in various analyses. Several studies have concluded that gas canal formation is essentially material- and geometry-specific [7–9]. From a comprehensive geometric and material perspective, this is correct, which is evident in the reference values of the various geometry groups that were listed previously. The quantitative effect of the partial filling degree, melt and tool temperatures, gas pressure, and other factors strongly depends on the available flow cross section that interacts with the material parameters. Nevertheless, the essential correlation between the process parameters and the gas canal cross section can be summarized as follows [6,10–13]:

- An increase of the cross section/reduction of the RWT results from increased melt temperature, reduced delay time, increased gas pressure, and shorter injection time during prefilling.
- Increasing tool wall temperatures lead to lower RWTs in thin-walled molded parts—however, the influence of the wall temperature is weak.
- The process dynamics are highly significant in gas injection. An increasing gas pressure induces an increasing formation speed for hollow spaces and is connected to an increased flow front speed.

The effects of changes in the gas pressure on the flow front speed verify the last bullet point (see Figure 3.5) [14,15]. Increased process dynamics are evident in the case of higher pressures. They are combined with a lower RWT.

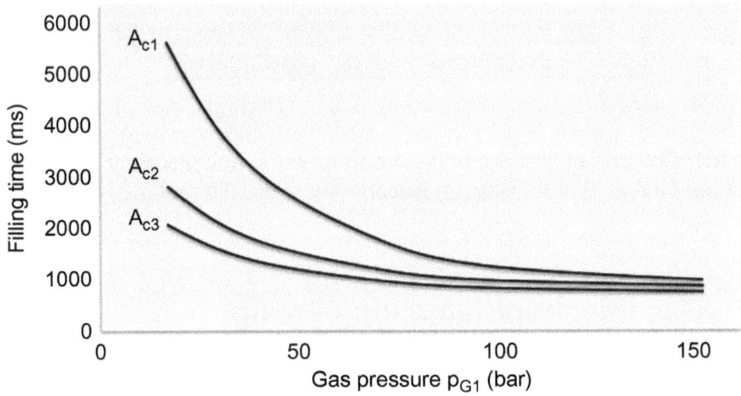

Figure 3.5 Schematic depiction of filling time over gas pressure (prefilling degree $\zeta = 80\%$) [13], $A_{c1} \approx 70\ mm^2$, $A_{c2} \approx 170\ mm^2$, $A_{c3} \approx 270\ mm^2$, material: SB.

As tests have shown, an increasing gas volume stream leads to clearly altered RWTs, despite the fact that the setting for the gas pressure does not change [16]. The increased gas bubble expansion speed and the higher shear speed lead to an enlargement of the cross section of the hollow space [17].

The parameters melt temperature and delay time also cause the process dynamics to change by altering the melt viscosity [13,18]. In Figure 3.6, the influence of the gas pressure on the cross section

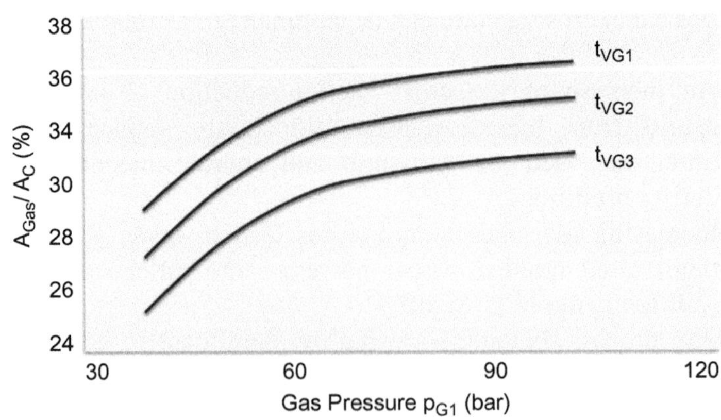

Figure 3.6 Schematic depiction of the gas bubble cross section (here normed to A_c), the gas pressure p_{G1} and the delay time t_{VG} ($t_{VG1} = 0$ s, $t_{VG2} = 0{,}3$s, $t_{VG3} = 0{,}6$s) [13].

A_{Gas} (divided through A_c) is depicted for a thin-walled component with gas-guiding ribbing ($A_c \approx 70\,mm^2$). A reduction of the delay time results in hollow spaces with larger cross sections.

Figure 3.7 depicts a typical curve of the internal pressure in the mold during the GIT process. A sensor inside the tool that situated near the gas injection point measures the pressure. The course of the curve provides information about several characteristic values:

- p_1—The maximum pressure is reached directly after completing partial filling; a delay time ensues, and the gas pressure decreases.
- p_2—The pressure in the molded part increases again after the injection of gas; p_2 sets the melt in motion.
- p_3—After a short pressure drop at p_3 due to the instantaneous hollow space enlargement and relatively slow tracking of the gas.
- p_4—After partial filling is completed, the pressure increases to the level set for the formation of a hollow space p_4.
- p_5—The gas-holding pressure was reached.
- p_6—As the freezing of the plastic progresses, the pressure slowly drops to p_6; subsequently, the holding pressure ends.

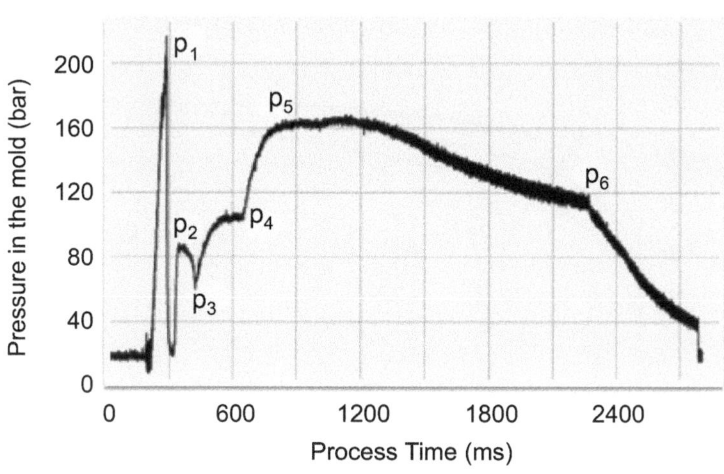

Figure 3.7 Typical internal tool pressure curve in a GIT injection molding tool with a internal pressure sensor close to the gas injection point [19].

Accordingly, the formation of a hollow space takes place between pressure steps p_2 and p_4.

As Figure 3.6 shows, when the gas pressure is sufficient, a threshold value $A_{Gas,limit}$ is reached for the gas bubble cross section, which cannot be increased by a further increase of the pressure. This threshold value, which represents the maximum achievable cross-section value, strongly depends upon the geometrical boundary conditions, the throughput characteristics of the gas injector, and the flow properties of the material.

In order to summarize the influences of all named process parameters, A_{Gas} is referred to as $A_{Gas,limit}$. This means that values determined for a gas pressure for one geometry under varying process parameters are referred to their maximum achievable hollow space cross section, $A_{Gas,limit}$. The correlation depicted in Figure 3.8 applies. As the pressure increases, the curve shown approaches a threshold cross section, which is typical for this situation. With regard to the typical threshold cross section, the influencing factors can change the initial increase of the curve. This means that the threshold cross section is reached at higher or lower pressure values due to the characteristic cross section, material, temperature, and other elements [13]. Determining the threshold cross section and providing an exact description of the influences of the parameters can be achieved experimentally for this case (i.e., the tool–machine combination in question).

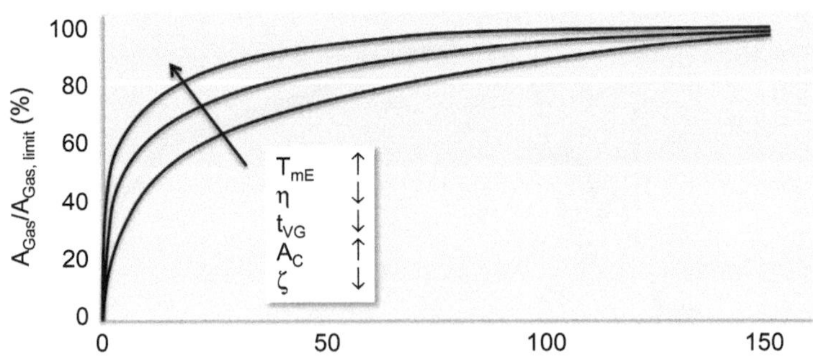

Figure 3.8 Geometrically comprehensive correlation between the gas pressure and relative gas bubble cross section $A_{Gas}/A_{Gas,limit}$, as well as other process parameters.

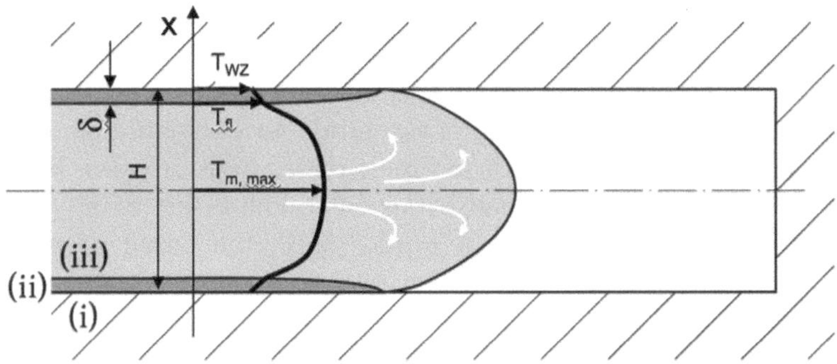

Figure 3.9 Schematic illustration of the temperature curve and the frozen marginal layer δ in the melt before injection of the expanding gas bubble (T_{WZ}, mold temperature; T_{fl}, flow temperature; $T_{m,\ max}$, maximum melt temperature; H, flow channel height).

In the abovementioned observations, the tool temperature was not included. It primarily influences the thickness of the frozen marginal layer (see Figure 3.9). The condition for transition between a frozen marginal layer δ and the melt at that point,

$$x = \frac{H}{2} - \delta, \tag{3.1}$$

is falling below the flow temperature T_{fl} in the frozen area:

$$x = \frac{H}{2} - \delta, \quad T = T_{fl}. \tag{3.2}$$

Common values for the frozen marginal layer equal 1/10 mm. According to L. S. Turng between the frozen marginal layer and the gas canal, there is always a melt phase [18]. Thus, the area available for the formation of a gas canal is strongly influenced by the tool temperature only in the case of gas-guiding geometries with small cross sections.

3.4.3 Cooling Phase/Formation of the Quality Features in the Component

The process parameter that sets GIT apart from standard injection molding is the gas-holding pressure p_{G2}. It is responsible for the same function that the holding pressure has in conventional injection molding (namely, to compensate shrinkage and reduce warpage).

The component being manufactured is pressed against the tool wall throughout the entire cooling time in the direct area of the gas canal. It cools quickly owing to a good transferral of heat caused by the high contact pressure [20]. In contrast to conventional injection molding, it is possible to cool the melt under high pressure both near the sprue and far from it when using GIT. The effect of p_{G2} is illustrated in the *p-v-T* diagram. In Figure 3.10, the curve of an amorphous material is depicted [13].

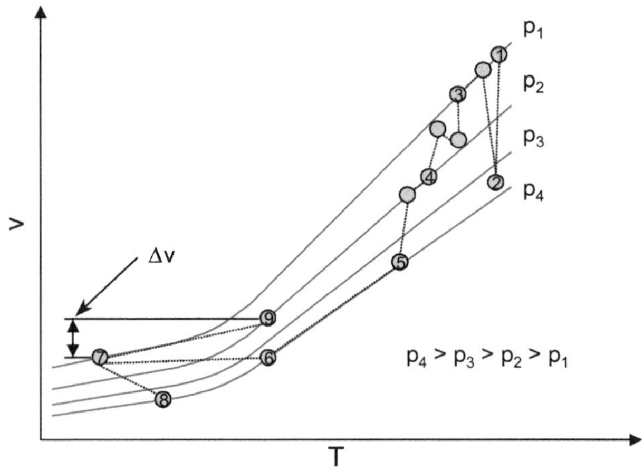

Figure 3.10 Schematic illustration of the process of GIT in a *p-v-T* diagram.

The GIT process consists of the following steps (with numeration as shown in Figure 3.10):

①–②—Injection of the melt.

②–③—Gas delay time (here: the pressure decreases rapidly after partial filling).

③–④—Blowing the melt and building pressure (here: the pressure declines after the injection of gas due to the movement of the melt, only to be built up again to the set level after the injection of the remaining filling).

Note: The actual cooling in the area between steps ① and ④ is quantitatively lower than shown in the graph. The time period available for cooling between steps ① and ④ is less than a second or up to a few seconds, and it results in very little cooling in the core of the observed molded part volume.

④, ⑤, ⑥—The first pressure step is held first. Subsequently, it is increased to the pressure level p_{G2} (pressure-holding phase).

⑥–⑦—The end of the pressure-holding phase and removal of the piece occur.

⑧, ⑨—Alternatives: Pressure holding with lower pressure ⑨ or higher pressure ⑧.

The advantage of GIT becomes clear in the cooling of the isobaric value from steps ⑤ to ⑥ at 300 bars. For comparison, cooling to the 100-bar line between steps ④ and ⑨ can be referenced. In this case, the glass transition temperature would be reached at a significantly higher volume. As a consequence, volume shrinkage depicted in the graph results and can lead to warpage and internal stresses in the component.

In the cooling phase after the formation of the hollow space, which starts at position ④, the RWT continues to cool. In Figure 3.11, the heat flows for the cooling are shown in the direction of the gas canal and of the tool. The contributions of the gas canal can be disregarded at first because heat dissipation is not possible after the initial absorption of heat by the gas. Therefore, complete cooling must be achieved via the tool wall. Thus, the melt of the RWT is cooled in a one-sided manner. For example, a strongly increased cycle time can be expected for the partially thick-walled component depicted in Figure 3.12 due to the approximately 1.5-mm-thick RWT, which is thicker than would be expected for a 2-mm-thick, thin-walled, adjacent area.

Cooling in the area of the RWT can be estimated using the Fourier heat conduction equation [6]. In a one-dimensional (1D) case and when

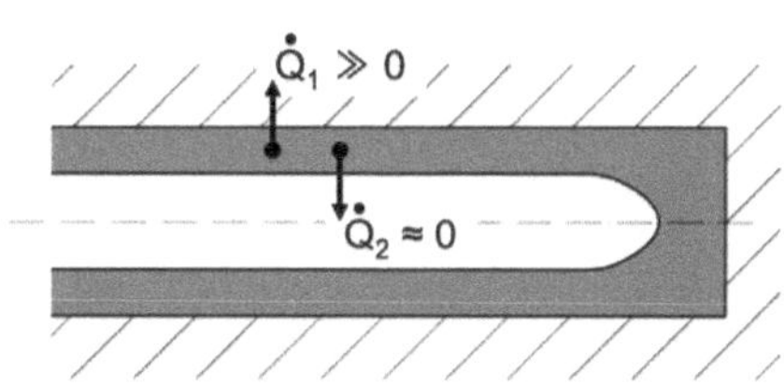

Figure 3.11 Schematic illustration of the heat flows in a GIT-molded part. $\dot{Q}1$ cooling in the direction of the tool and $\dot{Q}2$ cooling in the direction of the gas canal.

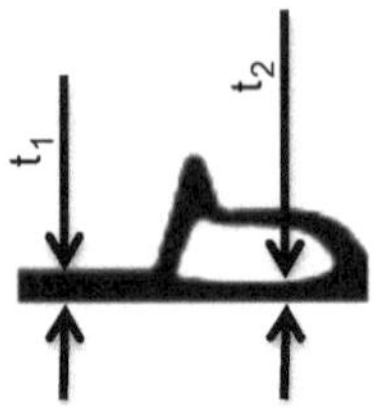

Figure 3.12 Exemplary illustration of a partially thick-walled component with $t_1 = 2$ mm wall thickness in the thin-walled area, and a RWT of approximately $t_2 = 1.5$ mm.

using the simplification of the constant heat conductivity λ, the equation is as follows:

$$\frac{\partial T}{\partial t} = \frac{\lambda}{\rho \cdot c_p} \cdot \frac{\partial^2 T}{\partial x^2}. \tag{3.3}$$

The following hypotheses are established:

1	The initial temperature in the plastic interface at point $x = 0$ is only slightly lower than the mass temperature T_M of the plastic at the point in time $t = 0$; it is referred to as T'_M.	$x = 0$, $t = 0$: $T = T'_M$
2	Simplified, isothermic conditions are set for $T(x, t = 0)$.	$T(x, t = 0) = T'_M$
3	A constant tool wall temperature T_W is set.	$x = d$, $t \geq 0$: $T = T_W$
4	For the solution of the differential equation, it is assumed that the geometry of the RWT is similar to the geometry of a plate.	

With a = the effective, invariant temperature conductivity, the following solution results for the time period after the formation of the hollow space (Pos. ④, $t = 0$) [6]:

$$T(x,t) = T_W + (T'_M - T_W) \cdot \frac{4}{\pi} \left[\sum_{n=1,3,5}^{\infty} \frac{1}{n} \cos\left(\frac{\pi \cdot n \cdot x}{2 \cdot d}\right) \exp\left(\frac{-a \cdot n^n \cdot \pi^2 \cdot t}{4 \cdot d^2}\right) (-1)^{\frac{n-1}{2}} \right]. \tag{3.4}$$

In Figure 3.13, the cooling curves calculated with Eq. (3.4) are depicted qualitatively for various cooling times or gas holding times t_{GH1} to t_{GH5}.

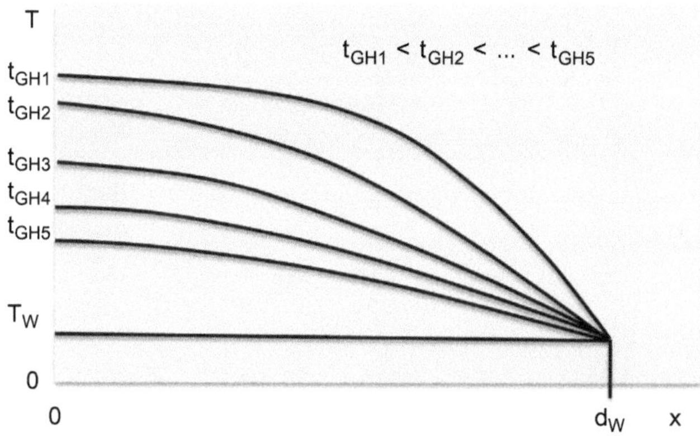

Figure 3.13 Schematic temperature curve in the area of the RWT (d_W) during cooling.

3.5 Influencing the Form and Position of the Gas Bubble

Basically, the outer contour of the gas-guiding geometry determines the form of the cross section of the gas bubble. Thus, a component with a circular cross section displays the most even wall thickness distribution. According to Moritzer's dissertation [6], it clearly shows the tendency of the gas bubble to take on the form of the cross section of the outer contour.

In the case of bar-shaped, thick-walled formed pieces, it is especially important to take into account that an uneven gas bubble cross section can form along the gas bubble for the following reasons:

- There is an orientation of the gas bubble toward the inside of the curve whenever the gas bubble must change direction. This effect is connected to a reduction of the gas bubble cross section. The position of the curve of the gas bubble is clearly visible in Figure 3.14 in the areas between MP 2 and 3, MP 4 and 5, MP 10 and 11, and MP 14 and 15.
- An expansion of the gas bubble (meaning an enlargement of the gas bubble cross section) is probably visible in this area due to a strong acceleration of the formation of the hollow space in the area that was filled last (the area that is not filled yet after partial filling) [6,14]. Figure 3.15 provides a schematic illustration of this effect.

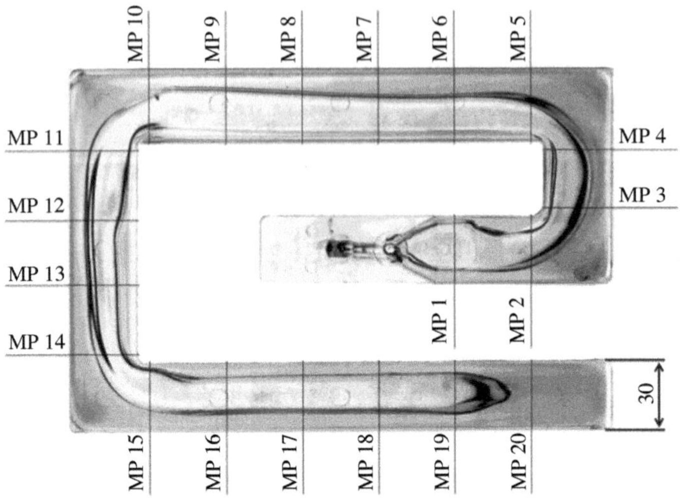

Figure 3.14 Illustration of the progress of a hollow space in a bar-shaped, thick-walled component. $A_c = 270\ mm^2$; material: transparent SB.

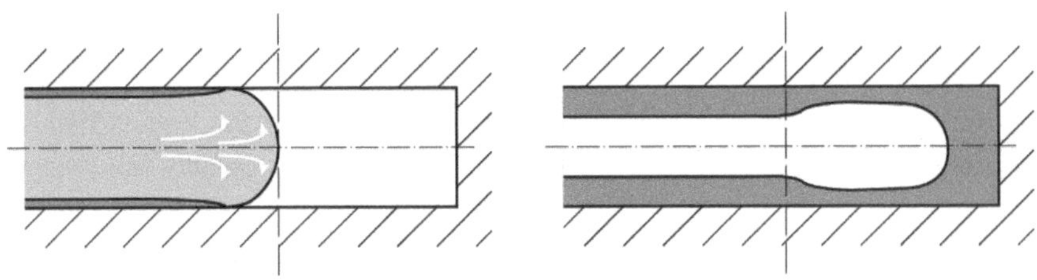

Figure 3.15 Schematic illustration of the expansion of a gas bubble in the last filled area. Left: partially filled mold; right: completely filled mold.

- When the outer contour of the gas-guiding geometry changes, the gas bubble follows the direction of the outer contour. A schematic depiction of this effect is shown in Figure 3.16.

Generally, the gas bubble is more stable in materials with higher viscosities than in materials that flow easily. One option to stabilize the position and cross section of the gas bubble by the process setting is to work with low temperatures and an increased delay time [19].

A gas breakthrough in the flow front and penetration of the gas into the thin-walled molded sections are typical defects of GIT that can occur in critical components (see Figures 3.17 and 3.18). Both effects (the finger effect and gas breakthrough) can have constructional causes (e.g., an incorrect layout of the tool) or may be attributed to parameter settings or parameter fluctuations [21,22]. If a gas breakthrough occurs,

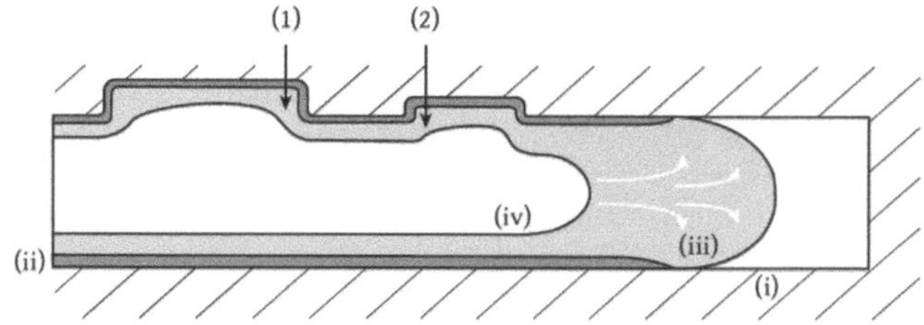

Figure 3.16 Schematic illustration of the alteration of the gas bubble cross section due to the outer contour of the gas-guiding geometry; (i) mold, (ii) frozen layer, (iii) melt, (iv) gas bubble, (1) and (2) melt accumulation/thick-walled area.

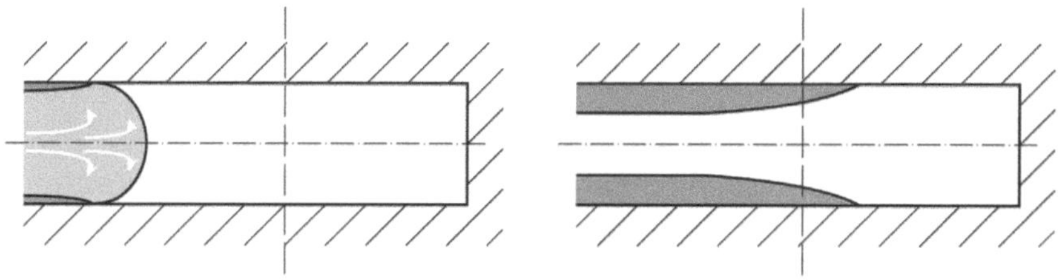

Figure 3.17 Gas breakthrough on the flow front.

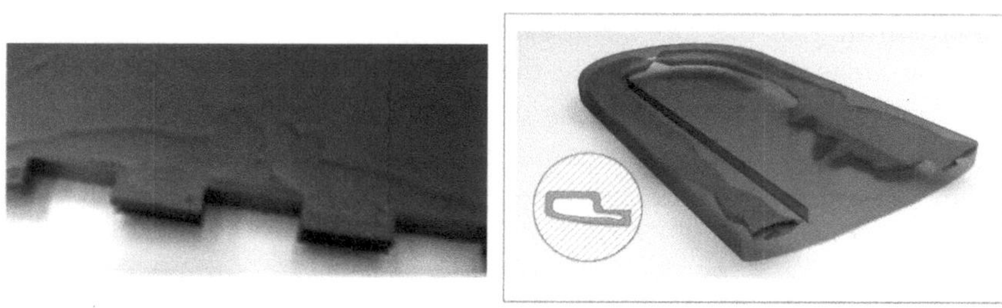

Figure 3.18 Image of a component with the finger effect (left); schematic illustration of the finger effect (right).

the component is no longer usable. If gas penetrates the thin-walled molded sections, the component does not necessarily become unusable but clear markings can form on the surface, which can reduce the optical quality of the component [23].

If the gas breakthrough or the finger effect can be traced to an insufficient molded part or tool layout—that is, incorrect position of the melt and gas injection, an unsuitable component geometry, a disadvantageous flow front course, etc.—a constructional change should be made (further information concerning this is provided in the last section of this chapter). If the formation of fingers occurs due to a shrinkage effect, changing the delay time is an option. Figure 3.19 shows the effect of extending the delay time.

Figure 3.19 Beginning finger effect (A, left) in comparison to an improvement as a result of extending the delay time (B, right).

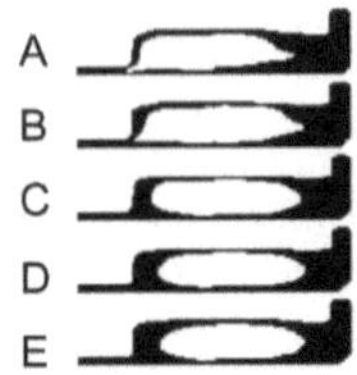

Figure 3.20 Example of the influence of the delay time on the position of the gas canal (partially thick-walled molded part; A, B, C, D, E—gradual increase of the delay time from 0.5 to 2.5 s).

Extending the delay time can also be used to center the position of the gas bubble. Figure 3.20 shows the effect of an extension of the delay time from 0.5 to 2.5 s in the case of a partially thick-walled component (material ABS, with dimensions of the thick-walled area of approximately 30 mm × 6 mm) [19].

The shifting of the gas canal can be explained by the cooling of the marginal areas that occurs due to the longer delay time. The temperature profile in the cross section of the components displays a rapid decrease near the tool wall, while a pronounced maximum is reached in the center of the cross section. As a consequence, the hollow space is centered [24].

3.6 Influencing the Surface Quality

Because the gas pressure builds up with a slight delay after switching from melt injection to gas injection, so-called changeover marks can occur during partial filling. The reason for this is the short-term standstill of the melt between injection and gas injection. If the machine controls enable it, it can help to start the injection of the gas before finishing the injection of the melt in order to avoid this standstill.

One of the advantages of GIT is the even distribution of pressure in the inside of the component. This causes shrinkage to take place from the inside out, meaning that the gas presses the melt from the hollow space out onto the surface of the tool. This leads to an improved heat transition from the plastic onto the metal surface of the tool. Consequently, the surface of the tool is modeled very well. The result is a good dimensional accuracy with very little warpage, as well as high surface quality. However, there are two serious disadvantages that result in the case of a component design that is not optimal:

- *Visible differences in surface quality in the transition area from the gas-guiding geometries and adjacent areas*— While the shrinkage takes place in the gas bubble from the inside out due to the internal pressure, the ratio of internal stresses on the surfaces and tensile internal stresses in the core that is typical for injection molded parts results in the adjacent area. This usually correlates with shrinkage in the direction of the core, which ultimately results in a sink mark on the surface (see Figure 3.21). A phenomenon that results for structured tool surfaces as a consequence of this is plastic abrasion in the area adjacent to the gas-guiding geometry. A contamination of the structure occurs, which is also visible in the form of a surface defect.
- *Visible effects on the surface in the case of uneven wall thicknesses caused by an uneven gas bubble formation*— The gas bubble is meant to push the melt out of the gas canal into the adjacent areas, and, thus, compensate for shrinkage [25,26]. For this purpose, p_{G2} is kept high. As a consequence, the amount of shrinkage reduces if the melt can flow sufficiently [27]. Compensating shrinkage makes it possible to create shrinkage spears in the area of the gas-guiding geometry, as shown in Figure 3.22 [11,18].

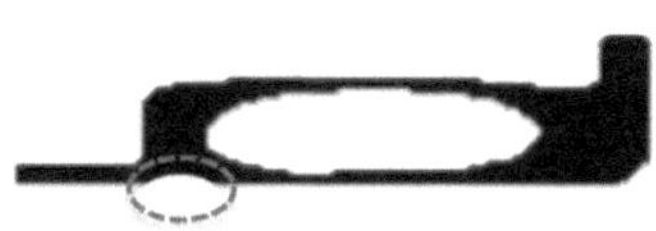

Figure 3.21 Schematic illustration of sink marks (see marked region) in the transition area between the gas-guiding geometry and adjacent areas.

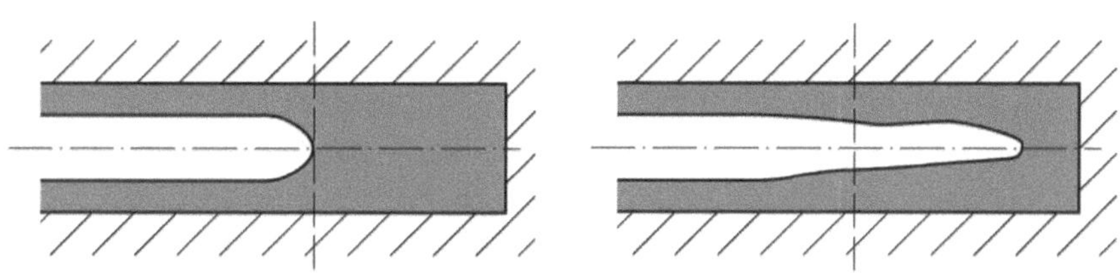

Figure 3.22 Illustration of the gas bubble formed during the volumetric filling (left) and the shrinkage spear that forms during the gas-holding phase (right).

In most cases, shrinkage spears are not critical. If they extend into the thin-walled area, they significantly influence the surface quality. Moreover, even if the latter is not the case, excessive shrinkage should be expected in the following cases:

- In the areas of the gas-guiding geometry, which were not blown using gas to create a hollow space
- In areas with uneven wall thicknesses
- In areas with an unevenly distributed gas canal [13,17,24,28]

When a typical gas-guiding geometry of GIT components is still in a compact state, it has a wall thickness that is not typical for injection molding components. This means that every area of the gas-guiding geometry that is not filled with a hollow space has a very high shrinkage potential. This can be compensated for to a certain degree by the shrinkage spear created by the gas bubble. In disadvantageous cases, the shrinkage (and accordingly also the warpage) are very high. In order to avoid this, the injection molding machine must accomplish an injection volume that is very reproducible from cycle to cycle. This is not a simple task for the plasticization aggregate due to the missing holding pressure phase. Special attention should be paid when using regrind materials, materials modified with lubricant, and thermoplastic elastomers and highly filled plastics.

3.7 Influence of Material Properties

In general, all thermoplastic materials are suitable for GIT. However, practical applications show that material changes, fluctuations in the charge, color changes, changes in the fiber and filling material content, and similar factors clearly affect the quality of the formation of a hollow space. These effects vary depending upon the application, ranging from light changes of the hollow space volume to finger effects and gas breakthroughs. The cause for insufficient quality caused by the material during GIT is generally a combination of a deficient molded-part layout and the material influences discussed next.

Generally, the thermal and the rheological material properties influence the gas canal cross section and the RWT. The section beneath the cross section of the gas bubble is generally referred to as the middle cross section, which is calculated by measuring several cross section areas of the hollow space along the gas-guiding geometry. Similar

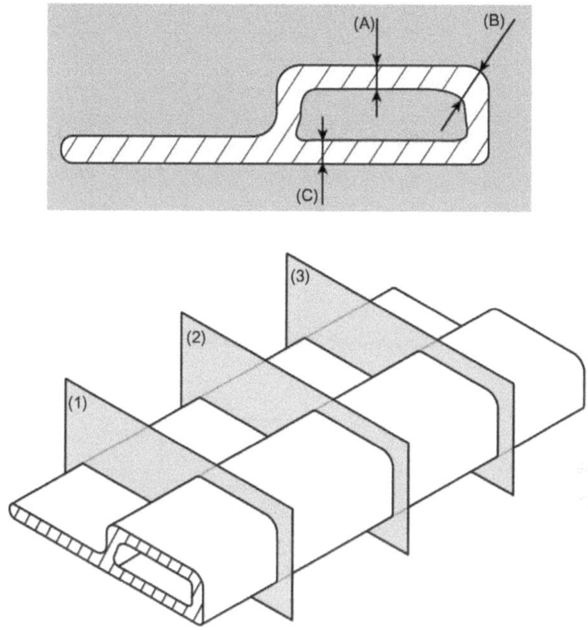

Figure 3.23 Illustration for the definition of the RWT.

conditions apply for the RWT. The measurement is carried out at several sections along the gas-guiding geometry (see Figure 3.23). The measured values are averaged so that the RWT is described in the following equation:

$$\mathrm{RWT} = \frac{\sum_1^{ij} \mathrm{RWT}_{ij}}{i+j}. \tag{3.5}$$

The thermal material properties can be characterized by the temperature conductivity a, which is calculated using the heat conductivity λ, the density ρ, and the specific heat capacity c_p. Figure 3.24 shows the values for various thermoplasts in the dependency of the temperature.

In the GIT process, materials with greater temperature conductivity tend to display higher values for the RWT because improved cooling results in larger, frozen marginal layers [29]. However, in practice, the temperature conductivity cannot be regarded separately from the rheological properties. Thus, a modification of the material with filling or reinforcement materials generally leads to an increase in heat conductivity; simultaneously, the rheological properties are changed greatly. Therefore, the effect of the RWT depends on the geometric influences. The question of whether the process is more thermally

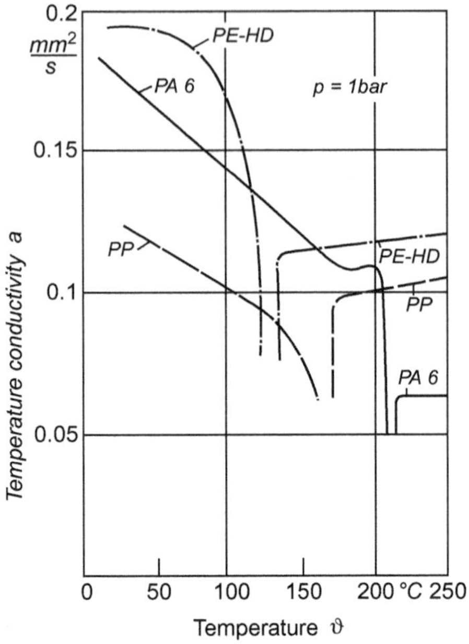

Figure 3.24 Illustration of the temperature conductivity of various thermoplasts.

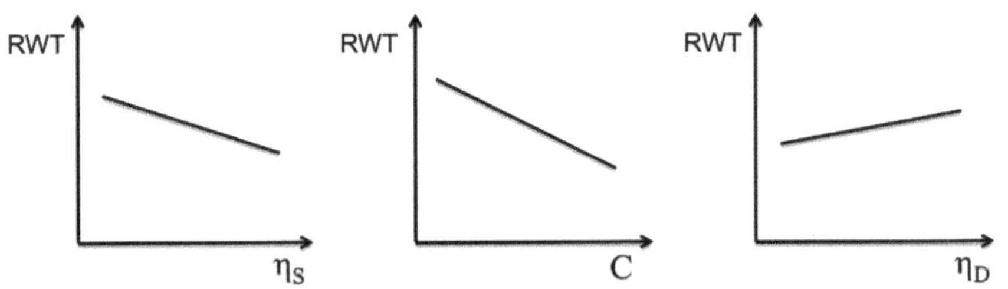

Figure 3.25 Schematic illustration of the rheological influences on the RWT.

dominated or whether the rheological properties are more predominant is raised. It can be said in response that the thermal properties play a lesser role.

In Figure 3.25, the influence on the RWT of the rheological material values shear viscosity η_S, elongational viscosity η_D, and the influence of the Carreau parameters C, which characterize the structural viscosity, is illustrated schematically. Next, the rheological influences are discussed individually.

3.7.1 Shear Viscosity

The shear viscosity changes due to the continuously changing shear speed during the expansion of the gas bubble. Accordingly, it is difficult to make a statement about the influence of the shear viscosity. In the research literature, varying (and sometimes even contradictory) statements about the effective direction of the RWT can be found. Yet, there is a consensus that the influence is very low, if it even can be observed [7,29,30]. An influence on the needed pressure for the formation of a hollow space is clearly visible. In Figure 3.26, the needed pressure for various employed materials is shown with the maximum achievable gas bubble cross section. The illustration is based on experiments performed on a component that belongs to the geometry group of thin-walled components with ribbing [6].

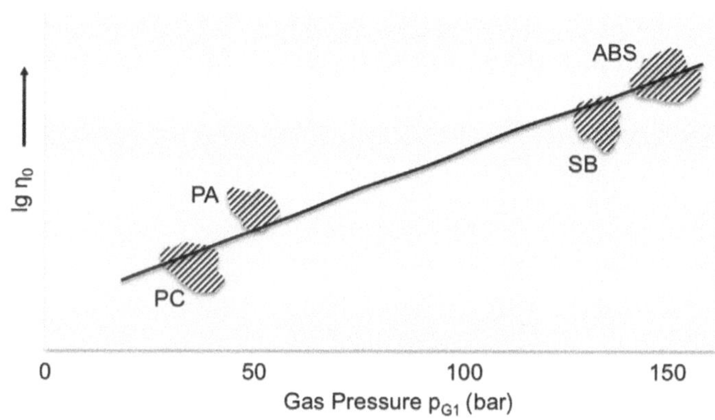

Figure 3.26 Illustration of the pressure needed for the formation of a hollow space for materials with varying zero viscosities η_0.

3.7.2 Shear Thinning

The shear thinning of plastics characterizes the deviation of the flow behavior from the Newtonic flow behavior, meaning that an increase in the shear speed $\dot{\gamma}$ does not manifest itself in a linear increase of the shear stress τ respective of the shearing forces in the melt flow, and, thus, create the pressure needed for the flow. Instead, the increase is less than proportionate. Thus, the resulting viscosity is not constant via $\dot{\gamma}$; instead, it decreases as the shear speed increases. This behavior can be described very accurately by the Carreau approach [31]:

$$\eta = \frac{A}{(1+B\cdot\dot{\gamma})^C}. \tag{3.6}$$

Parameter C describes the shear thinning; as parameter C and the shear speed increase, the viscosity increasingly deviates from the Newtonic flow behavior. An example of the derivation of the Carreau parameters from the viscosity curve is shown in Figure 3.27.

It was verified that increasing values for C correlate with decreasing RWTs in GIT [29]. An explanation can be found when observing the flow conditions in the flow canal or the injection molding tool [30]. Near the tool wall, next to the frozen marginal layer, a superelevation of the shear speed is available (see course $\dot{\gamma}$ in Figure 3.28) due to the fountain flow typical of highly viscose melts, which stick to the walls (see the course of the flow speed v in Figure 3.28). A minimum viscosity was recorded near the frozen wall (as also shown in Figure 3.28).

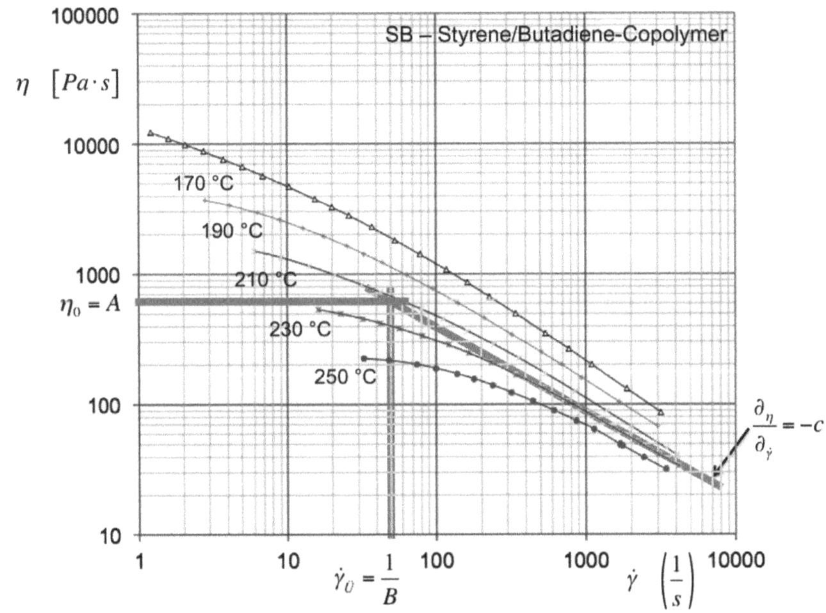

Figure 3.27 Illustration of the measurement values η over $\dot{\gamma}$ for SB at different tamperatures and labeling of the characteristic values A, B, and C of the Carreau approach.

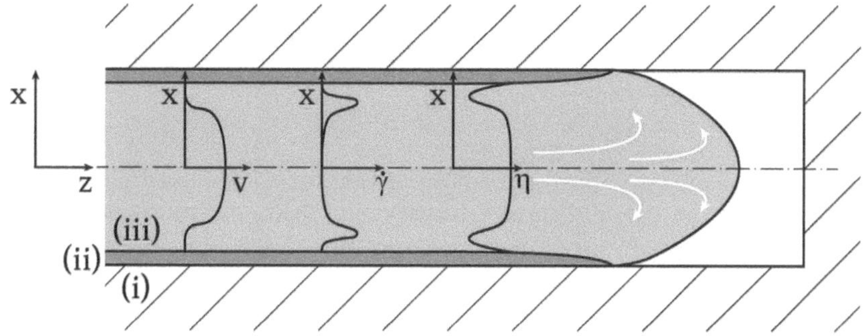

Figure 3.28 Schematic illustration of the courses of v, $\dot{\gamma}$, and η in the flow canal of an injection molding tool for melts that stick to the walls.

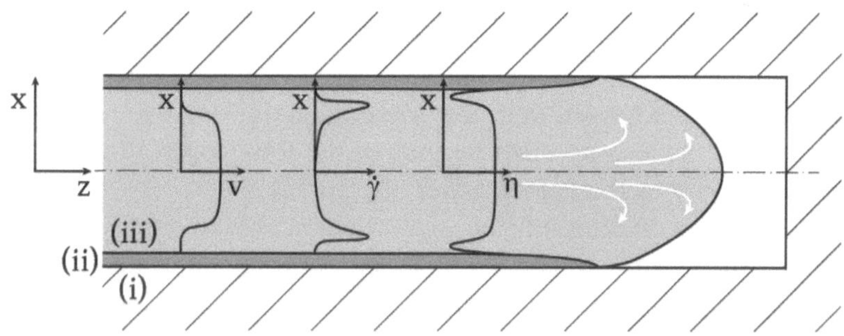

Figure 3.29 Schematic illustration of the altered courses of v, $\dot{\gamma}$, and η in the flow canal of an injection molding tool for melts that stick to the walls due to larger parameter C in the Carreau approach.

When the Carreau parameter C increases, another increase of the shear speed occurs, and thus a lower minimum viscosity also is manifest. Overall, a stronger, block-shaped flow profile results (see Figure 3.29 in comparison to 3.28). A more pronounced block flow (ceteris paribus) leads to a lower RWT.

3.7.3 Elongational Viscosity

The elongational viscosity η_D is a measure of the flow behavior that results from elastic deformations—that is, in the case of cross-sectional changes in the flow canal or on the flow front of the headstream during injection molding. The value of η_D depends on the strain rate and is defined as follows:

$$\eta_D = \sigma_D \dot{\varepsilon}. \tag{3.7}$$

An increase in elongational viscosity increases the elongation forces on the flow front, reduces the medium flow velocity, and ultimately results in the enlargement of the RWT.

3.7.4 Influences of Filling/Reinforcement Materials

In many cases, plastics are filled with filling materials like glass fibers, talcum, and minerals, in order to modify the mechanical properties. In the case of conventional gas-guiding geometry sizes, it can be assumed that the RWT reduces when filling materials are added [5,7,32]. If the process is thermally dominated, which is primarily the case for wall thicknesses that are approximately less than or equal to 2.5 mm, then it is also possible to enlarge the RWT owing to the greater heat conductivity of the filling materials [30,32].

An evident reduction of the wall thickness can be observed for materials filled with glass fibers as the content of fibers increases. In conventional GIT geometries, the reduction can equal up to 50%. High glass fiber content and long fiber reinforcements alter the structural viscosity of the material significantly (the Carreau parameter C rises) [30]. The altered flow behavior affects the process dynamics and leads to a smaller RWT.

3.8 GIT Tools and Machine Technology

A typical setup for GIT consists of the following components:

- Gas source (normally N_2)
- Pressure generator
- Gas control module
- Gas injector
- Pressure valves and connecting elements

Figure 3.30 shows a typical machine setup for gas-assisted injection molding.

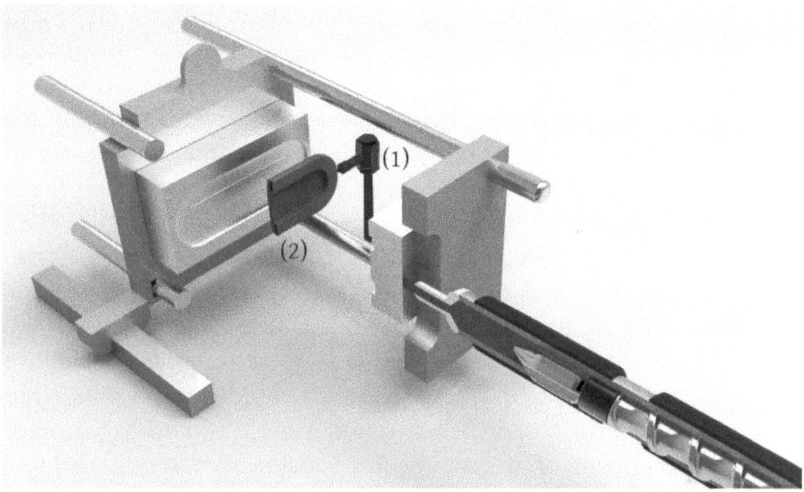

Figure 3.30 Machine setup for GIT, schematically: (1) gas injection; (2) molded part, with gas canals (sectional view).

3.8.1 Gas Source—Pressure Generation—Gas Control

The gas source can even be a simple gas flask. More complex systems employ gas-generating units that extract nitrogen from the surrounding air via a membrane.

In order to provide sufficient amounts of nitrogen with high pressure, an according compressor is needed. Conventional compressors provide initial pressures of up to 400 bars. Large equipment can provide up to 800 L/min. of nitrogen.

Gas control modules control the injection of the gas. The pressure/time profiles for gas injection can normally be programmed as desired. Proportional valves enable the precise adherence to the predefined pressure steps. Normally, a connection to the injection molding machine is established via the EUROMAP interface if an external control unit of the control module is used. Alternatively, the control module can also be integrated into the machine control of the injection molding machine.

3.8.2 Gas Injector

The injection of gas can take place either via the machine nozzle on the plasticization unit or directly in the tool via the hot canals or special tool nozzles. This procedure significantly limits the scope of action regarding the rheological layout of the tool, but it does have the advantage that only one marking is visible on the molded part.

In the case of thick-walled, bar-shaped components or simple, partially thick-walled geometries, the machine nozzle with a gas injector can be employed easily.

The injection of gas via canals is a more flexible option due to the ability to position the gas nozzle at various melt injection points more freely. However, the layout is also restricted to cases in which the melt and gas can be injected in the same spots.

Injection of the gas directly into the tool is advantageous because the positions for the melt and gas injections can be selected independent of one another. This creates a larger scope of action concerning the layout of the component and enables the design of complex gas-guiding canals. Especially in the case of thin-walled molded parts with ribbing, this variant of tool injector with a machine nozzle should be favored.

The simplest form of a gas injector located directly in the tool is drilling in the tool surface into which a cylinder equipped with a core is inserted, creating a ring gap with a gap width of less than or equal to 3/100 mm. A commercial ejector sleeve with an ejector is suitable for this purpose. It is designed such that both elements protrude into the tool cavity (see Figure 3.31). The ring cannot be too large, in order to prevent the melt from entering the gas canal. This is a very simple design, and it is sufficient from a process-technological standpoint, but it does have two, very important disadvantages:

- The removal of the injector is very complicated during maintenance.
- The throughput is relatively low due to the low gap width.

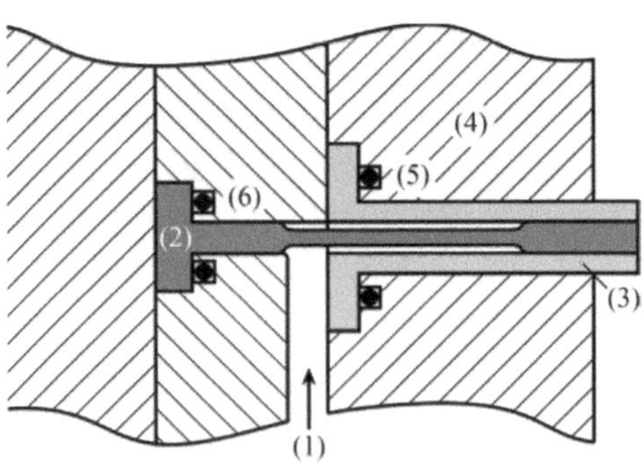

Figure 3.31 A drawing of the principle of a simple injector that was manufactured with standard components; (1) gas input, (2) ejector, (3) ejector sleeve, (4) mold wall, (5, 6) sealing rings.

In order to remedy these disadvantages, diverse variants of gas injections have been developed, which can be screwed into the tool surface and are therefore easily accessible when the tool undergoes maintenance. They are offered as prefabricated injector components by various suppliers.

In order to increase the volume flow in the gas injector, mobile injectors are offered in which a gas needle seals the injector until the melt has formed a frozen marginal layer on the injector and has shrunken. There are two functional variants:

- Pulling gas needle: By pulling the gas needle, the path for the gas is opened, creating a large opening in the tool.
- Piercing gas needle: By piercing the molded part, a large ring gap is created for the gas flow.

Both variants can be operated either externally (hydraulic or pneumatic) or by the action of a spring. The latter opens due to the pending gas pressure. The function of the stationary and pulling injectors requires the frozen marginal layer that has shrunken onto the injector to be broken through. Normally, this is not problematic. In contrast, the piercing gas needles open the component.

3.9 Design of GIT Components and Tools

Statements about the constructive design of molded parts can be found in various publications [5,6,9,17,23] in which different approaches are used to illustrate how to design GIT components correctly. In the following sections of this chapter, the most important information for the design of GIT components and tools has been summarized and divided into the following two aspects:

- Determination of melt and gas injection points
- Designing gas-guiding geometries

3.9.1 Determination of Melt and Gas Injection Points

The course of the flow front created by the positioning of the melt injection and the position of the gas injection must be coordinated with one another. The so called last filled area—concerning the melt

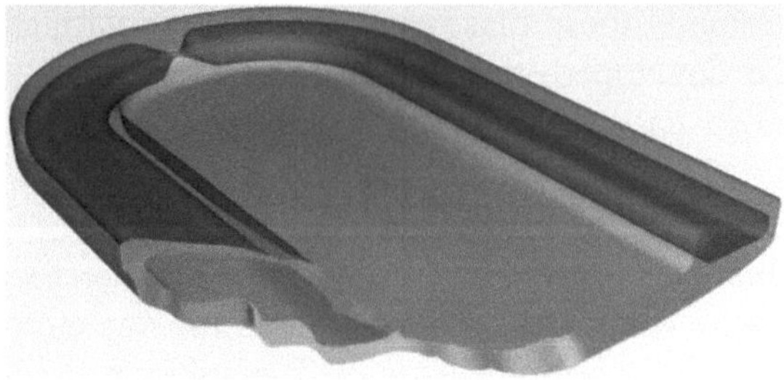

Figure 3.32 Illustration of the effect of gas being blown through the flow front.

injection—with the rest is decisive here. Since the gas always takes the path of least resistance when spreading in the cavity, it will always move into the not-yet-filled spaces when injected toward the melt to be displaced. Thus, it is advisable to position the injector so that the gas-guiding geometry matches the path of least resistance. The schematic illustration in Figure 3.32 provides an example of an unsuccessful design. After partial filling, the unfilled area is not at the end of the gas-guiding geometry, which inevitably leads to a displacement of the melt from the gas-guiding geometry into the thin-walled area. Large portions of the gas-guiding geometry remain solid, and the required volume for the remaining filling cannot be displaced in the thin-walled area. As a result, gas is blown through the flow front and an incomplete component is created (this and more examples in Table 3.4).

Essentially, the following can be said for the positioning of the gas injection point:

- If possible, gas should be injected directly into the thick-walled area and, thus, into the gas-guiding canal.
- The gas injection should be positioned on the end opposite the gas-guiding canal.
- In the case of gas spreading in two directions (branching), a defined mass template must be available to all branches because the flow resistance decreases as the flow path increases (unlike in conventional injection molding, where it becomes larger).
- The direction in which the injected gas flows should follow the same direction in which the melt flows.

Table 3.4 Examples of the Determination of Melt and Gas Injection Points [36]

Partially thick-walled component (schematic)	Partially thick-walled component (filling simulation)
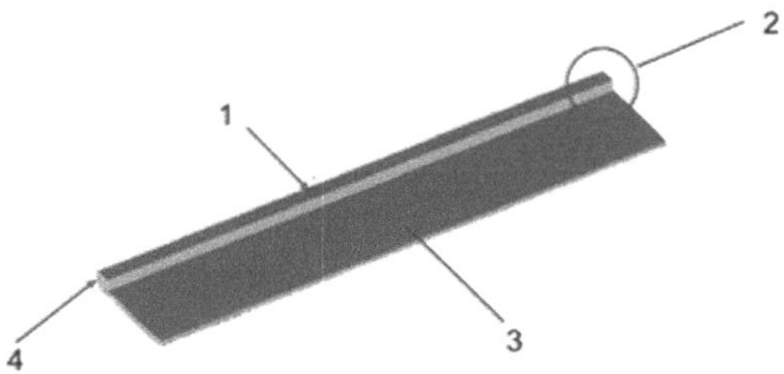 1 Thick walled area 2 Desired last filled area 3 Thin walled area 4 Gas injection Explanations: – Last filled area has to be at the end of the thick-walled area	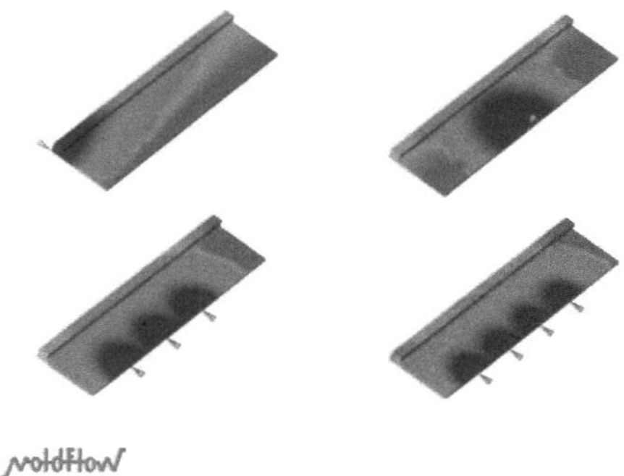 Explanations: – Different positions for melt injection lead to varying result concerning last filled area – Example 4 is the most expensive solution, but it also leads to the desired last filled area

Table 3.4 Examples of the Determination of Melt and Gas Injection Points [36]—cont'd

Thin-walled component with ribbing (schematic)	Thin-walled component with ribbing (filling simulation)
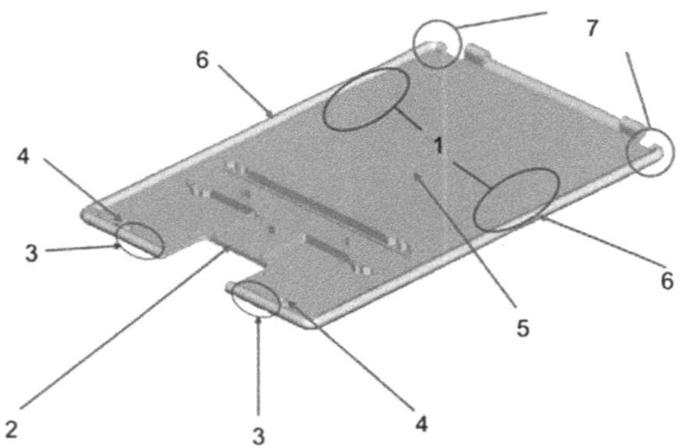 1 Finger effects 2 Melt injection 3 Shrinkage 4 Gas injection points 5 Thin-walled area 6 Thick-walled area 7 Desired last filled area Explanations: – Problem: Part shows finger effects and sink marks during series production – Gas injection points had to be moved in order to improve the development of the gas canal	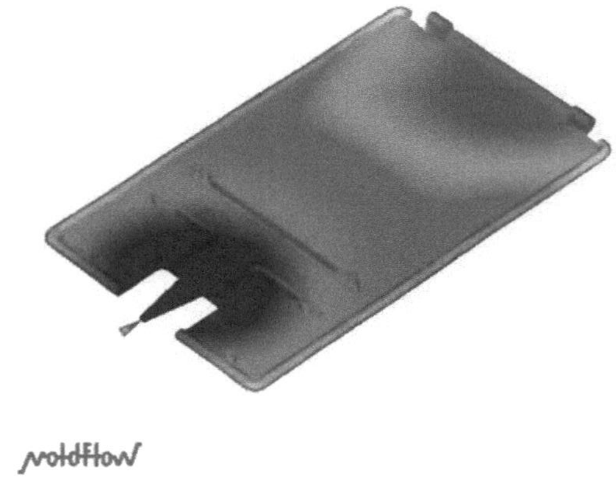 Explanations: – Melt injection point is not suitable to reach desired last filled areas – Due to other reasons in series production there was no way to move melt injection point – Finger effects are unavoidable

Bar-shaped, thick-walled component (schematic)	Bar-shaped, thick-walled component (filling simulation)
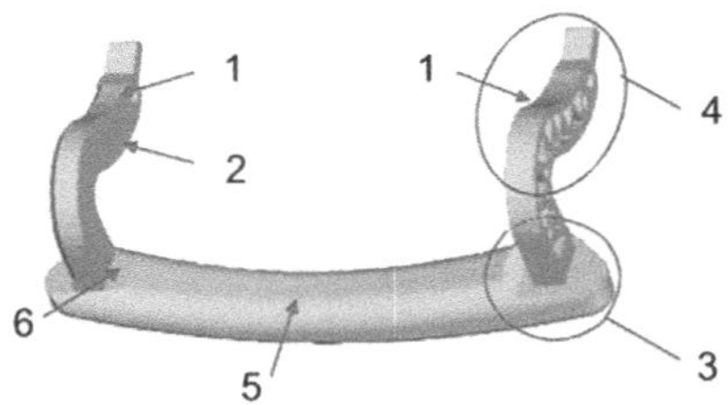 1 Should not be hollowed out 2 Melt injection 3 Desired last filled area 4 Realistic last filled area 5 Thick-walled area with gas canal 6 Gas injection Explanations: – Last filled area is possible at the end of the flow path only if one melt injection point is available	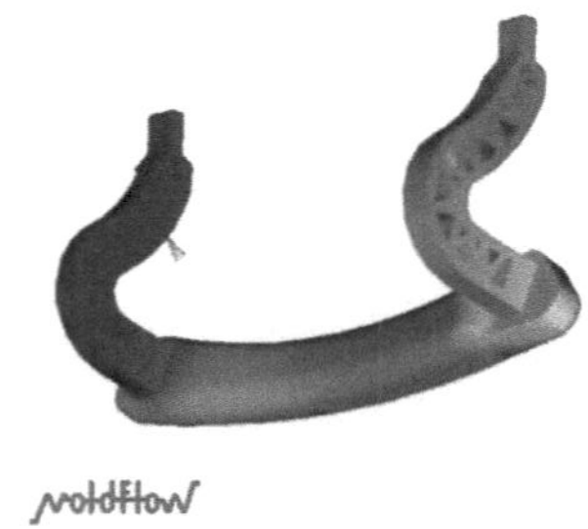

Table 3.4 Examples of the Determination of Melt and Gas Injection Points [36]—cont'd

Partially thick-walled molded part with a multidirectional gas-guiding canal (schematic)	Partially thick-walled molded part with a multidirectional gas-guiding canal (filling simulation)
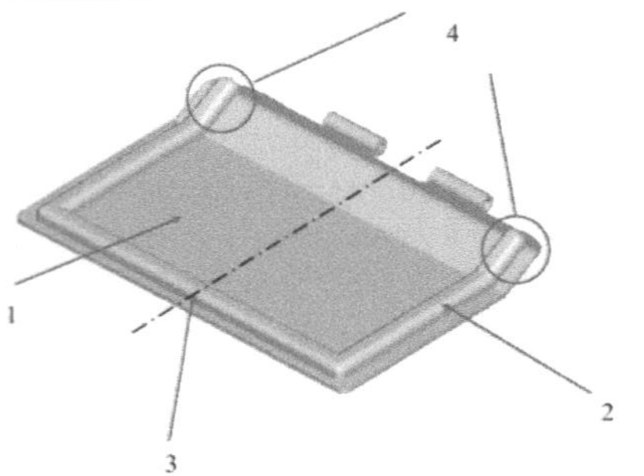 1 Thin-walled area 2 Thick-walled area with gas canal 3 Gas injection 4 Desired last filled area Explanations: – Two areas that are not filled completely after melt injection – One gas injection point	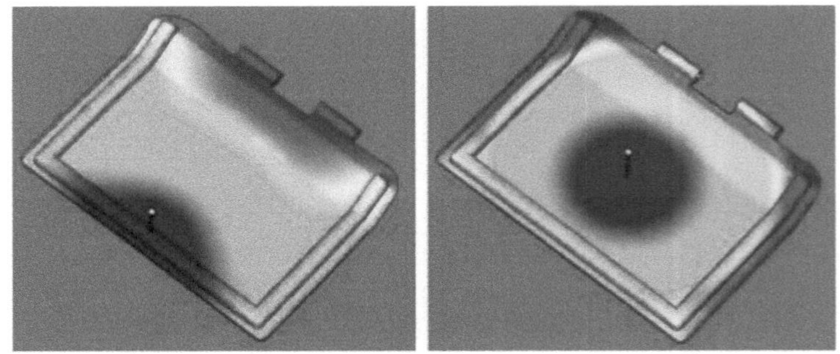 moldflow Explanations: – Melt injection in the center of the part in order to get a completely filled thin-walled area

3.9.2 Designing Gas-Guiding Geometries

A number of years ago, design guidelines for gas-guiding geometries were already compiled in various publications and dissertations [5,35]. In Table 3.5, the essential information for the design of thick-walled areas in a GIT component is summarized.

Table 3.5 Advantageous and Disadvantageous Designs of Gas-Guiding Geometries

	Disadvantageous, (1), (2)—mark points of interest	**Advantageous**
Connection of thin-walled areas to thick-walled areas of the molded part	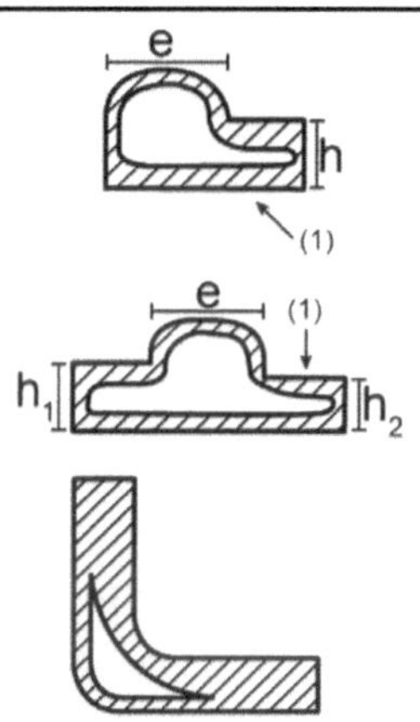	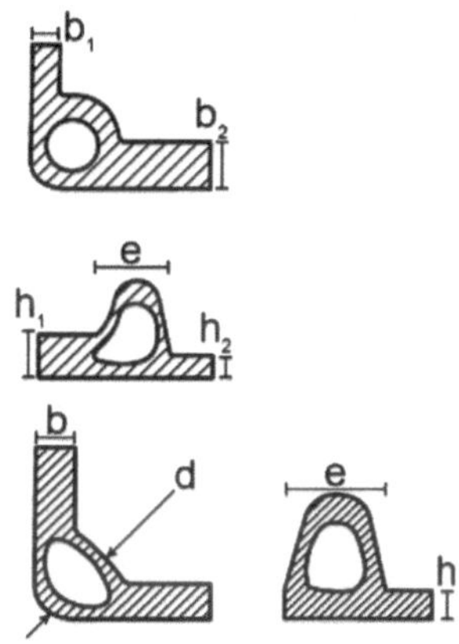
Transition from thick-walled to thin-walled area	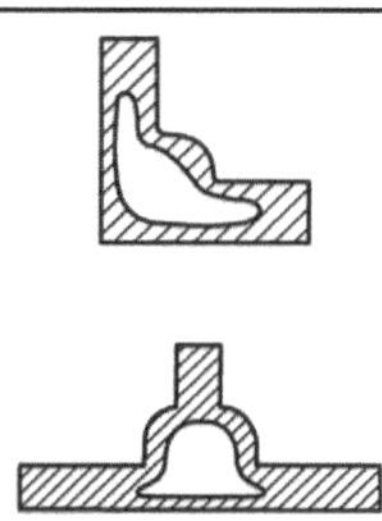	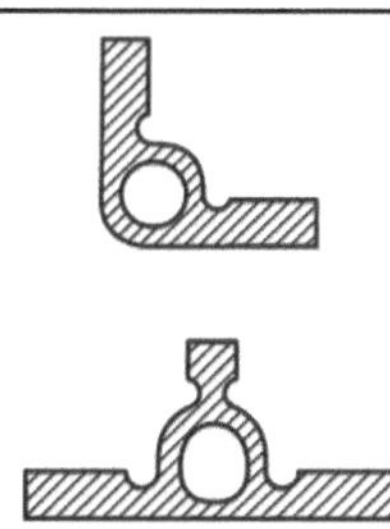
Cross sections of gas canal areas	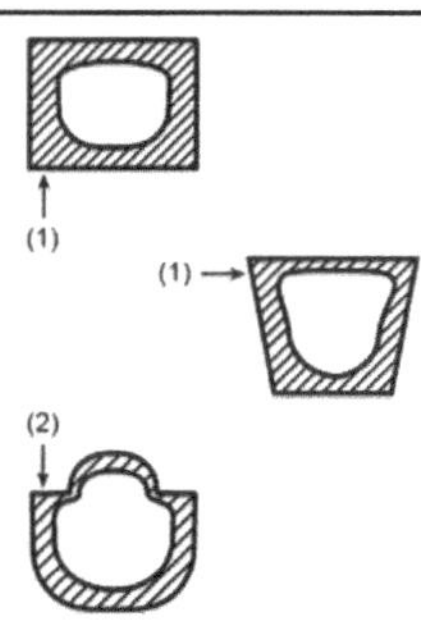	

Table 3.5 Advantageous and Disadvantageous Designs of Gas-Guiding Geometries—cont'd

	Disadvantageous, (1), (2)—mark points of interest	**Advantageous**
Width/height proportions	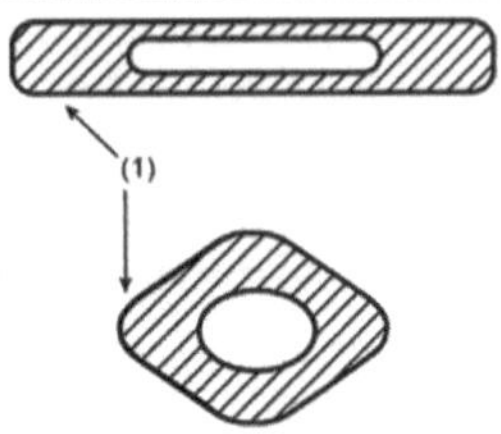	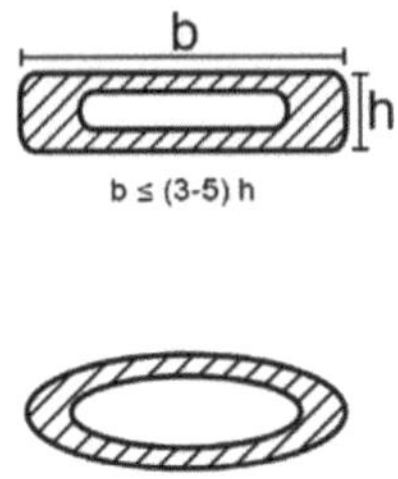
Length/ thickness proportions	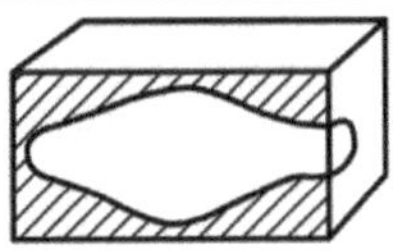	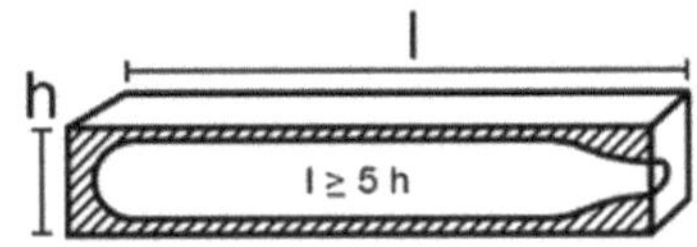
Influence of the curvature of the molded part	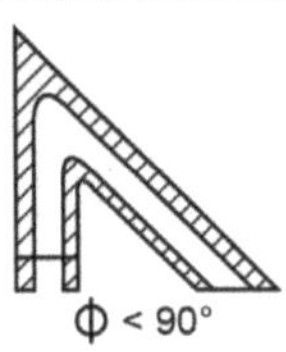	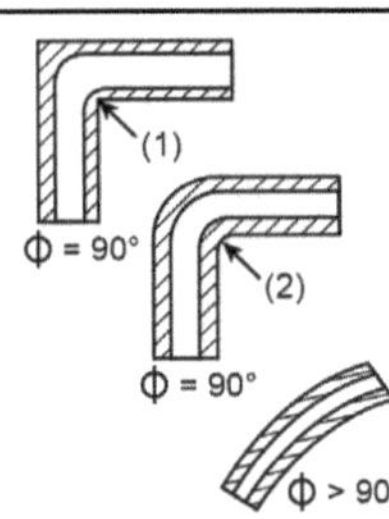
Design of cross section changes	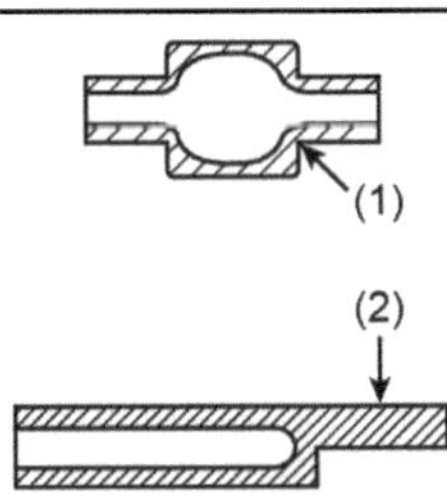	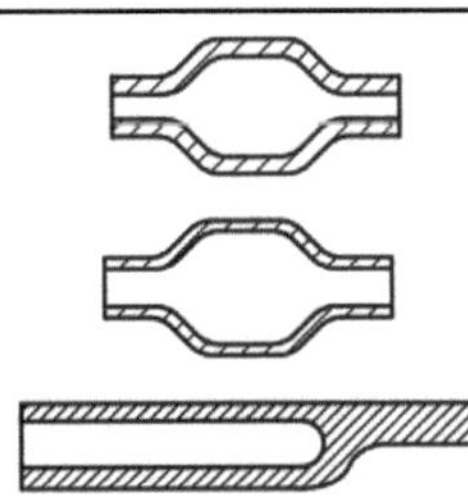
Branchings and gas canal junctions	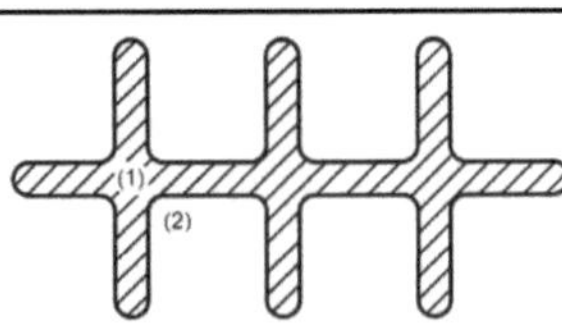	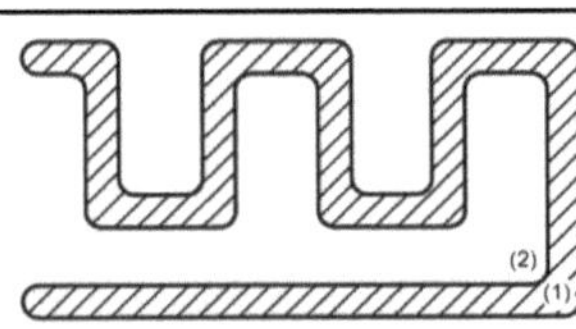

The descriptions in Table 3.5 are based on the dissertation by Ridder, and provide a very good overview of the dos and don'ts of gas canal design.

3.10 Variants of Fluid Injection Technology

In addition to the partial filling methods described in the previous sections, the side cavity method has become the most important variant of GIT. The side or overflow cavity method is described next. In addition, the following variants are described briefly:

- The push-back and core-pull technologies are rarely found in practice, but they are among the pioneer methods of GIT, so they are discussed briefly.
- Owing to its benefits in terms of its cycle time and larger hollow space cross sections, Water Injection Technology (WIT) is an interesting variant that is finding more and more users.
- If larger hollow spaces are needed, GIT can be combined with a blowing technology. This technology was published under the name GITBlow, and it is interesting when the goal is to combine large hollow-space volumes with thin-walled areas.
- The manufacture of multilayered hollow objects using GIT and WIT.
- Measures for the optimization of the surface.

3.10.1 GIT with a Overflow Cavity

In GIT with a side cavity, the component is completely filled before gas is injected. Additional volume is provided for the melt that is to be displaced. This additional volume—the side cavity—is closed by a slider at first. Directly after the tool cavity is filled completely, the slider and the side cavity are opened. The simultaneously injected gas displaces the melt into the side cavity. For this reason, the side cavity must be positioned at the end of the gas-guiding geometry. Figure 3.33 shows a sketch of this process.

The volume of the side cavity must equal the hollow space volume or the volume of the melt to be displaced. Therefore, it is advisable to

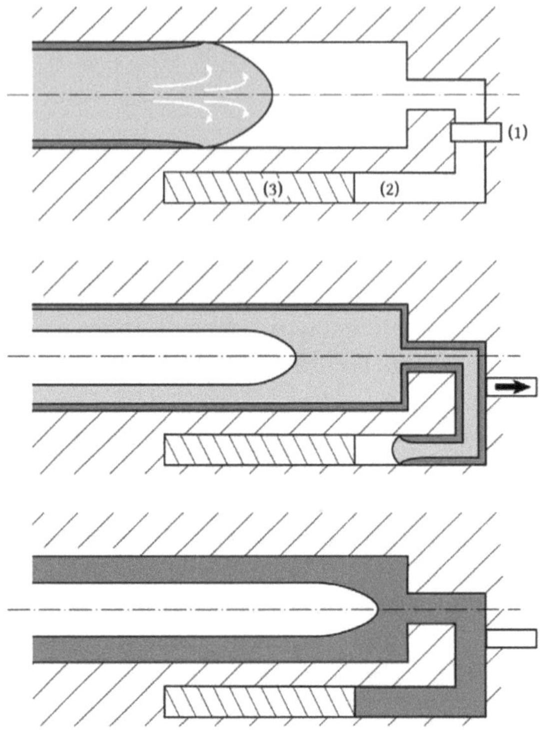

Figure 3.33 Schematic illustration of GIT with a side cavity. Top, before gas injection, closed slider; middle, during gas injection, open slider; bottom, after gas injection. (1) slider, (2) overflow canal, (3) adjustable size.

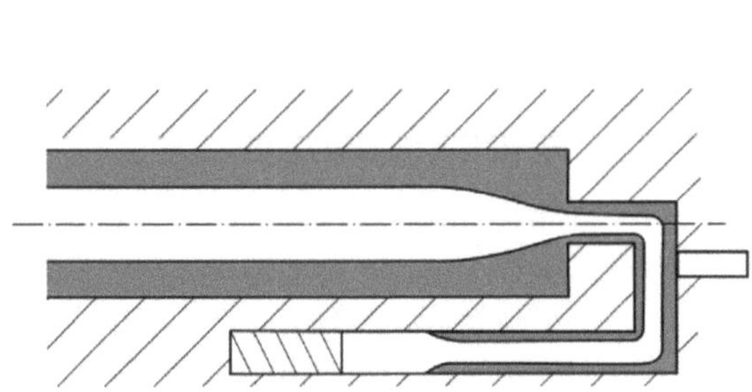

Figure 3.34 Schematic display of the larger side cavity.

have the option of setting the size of the side cavity by an exchangeable application. Figure 3.34 shows a representation of the mold after the injection of gas with an opened slider. The volume of the side cavity is too large, so the gas enters the side cavity.

The overflow cavity method has two obvious disadvantages:

- The side cavity must be removed mechanically after the manufacture of the parts.
- In addition to the visible gas position, the marking of the closed-off canal of the side cavity can be seen.

In contrast, the following aspects are advantageous:

- The filling of the molded part and the creation of a gas canal can be performed separately.
- The side cavity method enables a holding pressure before gas is injected, which makes it possible to influence the shrinkage and warpage in the thin-walled component area (without a gas canal).
- The method offers a large amount of design freedom concerning the gas-guiding geometry, regardless of the design of the rest of the component.
- The side cavity method makes it relatively easy to create several gas canals and branch the canals in the component.
- Several side cavities can be opened either at the same time or sequentially in order to control the melt flow.

Figure 3.35 shows an example of a component with two gas injection points and two gas-guiding geometries, of which one branches off. In this component, the side cavities 1 and 2 are opened directly after the

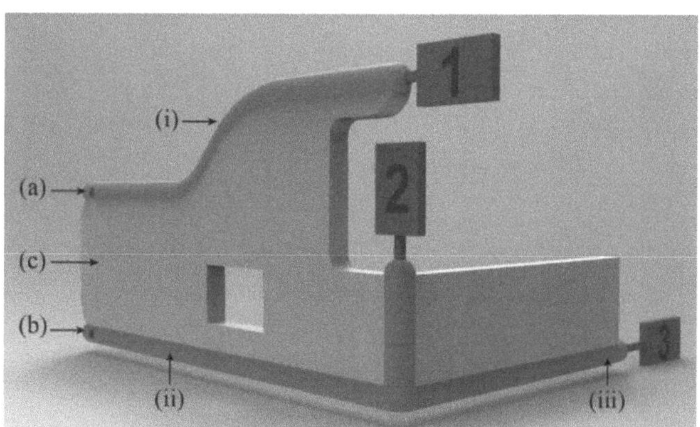

Figure 3.35 Schematic illustration of an exemplary component with two gas-guiding geometries.

component is completely filled, and a holding pressure has been applied shortly. Gas is injected into positions (a) and (b). After gas canals (i) and (ii) have formed in the direction of side cavities 1 and 2, side cavity 3 is opened. The gas displaces the melt in the direction of the gas-guiding canal (iii).

The side cavity method is especially suitable for thin-walled components with gas-guiding ribbing, in which a reproducible filling of the thin-walled area must be ensured.

3.10.2 Push-Back and Core-Pull Methods

A technology that is similar to the overflow cavity method is the *push-back method.* The difference between the two lies in the fact that the melt is not displaced into an additionally created hollow space, but is instead pushed back into the injection unit.

In contrast, an additional hollow space is used in the *core-pull method.* During the injection of gas into the fully filled cavity, a core in the tool is unlocked and pulled back. The freed volume is filled with gas, and the hollow space is formed as a result.

3.10.3 Water Injection Technology

In WIT, water is injected into the liquid melt instead of gas. The resulting benefits include shorter cycle times and larger hollow space volumes with smaller RWTs [30,37–40].

These advantages can essentially be attributed to three causes that are due to the properties of the water in comparison to the nitrogen used in GIT:

- The ability of the water to absorb heat: heat capacity or storage capacity
- The transfer of heat from the plastic to the fluid: heat transition coefficient and heat conductivity.
- Controllable process dynamics: compressibility

3.10.3.1 The Influence of the Heat Capacity or Storage Capacity

Shorter cycle times are achieved by means of a better cooling effect of water as opposed to gas. Water can absorb a significantly greater amount of heat during the creation of the hollow space than gas can.

The ability to absorb heat can be described by the storage capacity as follows:

$$Q_{\max} = c_{\mathrm{p}} \cdot \rho_{\mathrm{p}} \cdot V \cdot \Delta T, \tag{3.8}$$

where

c_{p} = Specific heat capacity when processing pressure is available
ρ_{p} = Density when processing pressure is available
V = Volume of the medium
ΔT = Temperature distribution

Regarding the available volume, the specific energy density can be referenced as a measure for the storage capacity of the according medium. The following applies:

$$q_{\mathrm{V}} = \rho_{\mathrm{p}} \cdot c_{\mathrm{p}} \cdot \Delta T. \tag{3.9}$$

If in two observed cases (GIT with gas and WIT with water), the same gas canal volume V should contribute to cooling the melt by ΔT, and if estimated temperature invariant values are used for ρ_{p} and c_{p} in the temperature range around 200°C at a processing pressure of 300 bars, the following expression can be used to estimate the cooling effect of the related media:

$$\frac{q_V}{\Delta T} = \rho_{\mathrm{p}} \cdot c_{\mathrm{p}}. \tag{3.10}$$

The following estimated values for the properties of water (H_2O) in the case of WIT, nitrogen (N_2) in the case of GIT, illustrate why a better cooling effect can be expected for. As an example for the WIT or GIT components material the value for Styrene-Butadiene (SB) is calculated:

$$\rho_{\mathrm{p}} \cdot c_{\mathrm{p}} \cdot (\mathrm{H_2O}) \approx 865\,\mathrm{kg}/m^{-3} \cdot 4.4 \cdot \mathrm{kJ}/(\mathrm{kg} \cdot \mathrm{K})^{-1} \approx 3800\,\mathrm{kJ}/(\mathrm{m}^3 \cdot \mathrm{K})^{-1} \tag{3.11}$$

$$\rho_{\mathrm{p}} \cdot c_{\mathrm{p}}(\mathrm{N_2}) \approx 180\,\mathrm{kg}/m^{-3} \cdot 1.1 \cdot \mathrm{kJ}/(\mathrm{kg} \cdot K)^{-1} \approx 200\,\mathrm{kJ}/(m^3 \cdot K)^{-1} \tag{3.12}$$

$$\rho_p \cdot c_p(\text{SB}) \approx 940\,\text{kg}/m^{-3} \cdot 1.8 \cdot \text{kJ}/(\text{kg} \cdot K)^{-1} \approx 1700\,\text{kJ}/(m^3 \cdot K)^{-1}. \tag{3.13}$$

3.10.3.2 The Influence of the Heat Transition Coefficient and the Heat Conductivity

The relationship between the heat transition coefficient between plastic and water (α_W) or plastic and gas (α_G) and the heat conductivity (λ) of the plastic is decisive concerning the question of whether a pronounced temperature gradient results in the area of transition. This relationship can be described by the dimensionless Biot number *Bi*:

$$Bi = \frac{\alpha \cdot L}{\lambda}. \tag{3.14}$$

L = characteristic length of the heat conductivity of a solid object

In the case of large Biot numbers, the temperature gradient in fluid is small in contrast to that in a solid object. If *Bi* is small, the temperature gradient in a solid object is relatively small as well.

When applied to fluid injection technology, this means the following:

- In GIT, α_G is in the following range of values:

$$\alpha_W = 10 \rightleftarrows 250 \frac{\text{W}}{\text{m}^3 \cdot \text{K}}$$

- In WIT, α_G is clearly smaller than α_W, with values in the following range:

$$\alpha_W = 600 \rightleftarrows 12{,}000 \frac{\text{W}}{\text{m}^3 \cdot \text{K}}$$

- In both cases, the flows are forced ($Re > 2300$). Thus, turbulent flows are available. Therefore, high values can be assumed for the abovementioned areas for α.

Consequently, it can be assumed that the temperature gradient in plastic in WIT is higher, and that the water leads to a cooler layer situated on the inside, which stabilizes the flow front (see Figure 3.36). The experimental results show that lower RWTs occur in WIT than in GIT, but the distribution of the wall thickness is also more even. This, in turn, leads to evener cooling and shrinkage, as well as warpage [30,38,40,41].

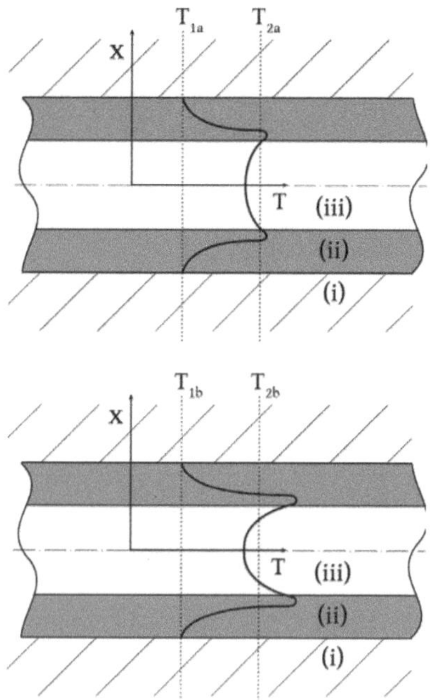

Figure 3.36 Schematic illustration of the temperature gradients in the RWT for a small *Bi* (top) and a large *Bi* (bottom); $T_{1a} = T_{1b}$, tool wall temperature; $T_{2a} \neq T_{2b}$, marginal layer temperature between the remaining plastic wall and the fluid. (i) Tool wall, (ii) remaining plastic wall, (iii) gas/water canal.

3.10.3.3 The Influence of the Compressibility of the Fluid

As is known, the compressibility of nitrogen is significantly greater than that of liquid water. Sufficiently high pressure during fluid injection can keep water permanently below the steam pressure curve, which stays far below 100 bars in the case of conventional processing temperatures up to 300°C. Thus, evaporation can be avoided if the pressure is set at a sufficiently high level.

Due to the low compressibility of water, two method variations of WIT can be employed [30]:

a. *Pressure-controlled p-*WIT, *which uses constant pressure for the injected fluid.* As in GIT, the volume flow (and, thus, the flow speed of the melt in *p*-WIT) moves in front of the driving fluid due to flow resistance. A speed gradient with strong acceleration of the flow-front speed in the direction of the flow path end occurs as a result.

b. $\dot{V}$-WIT *is controlled by the volume flow, which determines the volume flow for the fluid.* Due to a counter-pressure of the melt to be displaced, a need for pressure arises, which must be provided by the WIT facility. The advantage of $\dot{V}$-WIT is that no acceleration of the flow speed occurs. Thus, the dynamic conditions during fluid injection can be controlled better.

Whether, and to what extent, an advantage is given to the user in each case is not fully clear. Published experimental results show that slightly lower RWTs can be achieved using $\dot{V}$-WIT when very long flow paths are realized [30].

3.10.3.4 Systems Technology Requirements for WIT

Using water results in two essential differences between WIT and GIT:

- Water has a higher viscosity than gas, so it must be injected using larger injectors, or larger flow cross sections in the injector. For this reason, systems with active closing mechanisms are exclusively employed for WIT [42].
- The water must be removed from the component after injection.

Various concepts were developed to remove the water out of the component, and these will be covered later in this chapter [43].

Actions that can be taken in the tool include the following:

- The water is blown out with compressed air, either by introduction in one spot or by an alternating introduction of the compressed air.
- Adding blowing agents to the water results in an independent expulsion of the water after decompression.
- Water can also be removed by evaporation after decompression at temperatures above 100°C.

Outside of the tool, in some cases, appropriate handling of the component can enable the water to flow out of strongly curved components.

3.10.4 GITBlow (GIT Combined with Blow Molding)

The method variant known as GITBlow is based on a patent by this author; it was first published in 2005/2006 [44–46]. Additional publications followed, which focused on process control and component design [47–49], the distribution of the wall thickness prior to and after blowing [50], and temperature control in the gas canal prior to and during blowing [51,52]. In 2014, a dissertation about GITBlow was published [53].

GITBlow is a variant of fluid injection technology that is used to extend GIT by one additional step. After creating a gas bubble along the gas-guiding geometry, a hollow space is blown in a direction that diverges from this geometry. Thus, the method is similar to injection stretch blow molding. However, in terms of its original geometry (preform), it is not limited to hollow spaces that can be created with a core. Instead, it can use any hollow-space geometry that GIT can create for blowing. The GITBlow method enables a realization of molded parts with the following:

- A large hollow space cross section and a very small RWT
- An even distribution of the RWT

All three geometric groups conventionally used in GIT (i.e., thick-walled, bar-shaped, partially thick-walled, and thin-walled with gas-guiding canals) can also be used in GITBlow.

The general procedure involves the following steps:

1. Partial filling of the component (or, alternatively, a complete filling)
2. Performing the remaining filling and creation of the hollow space along the gas-guiding geometry using GIT (or, alternatively, using the side cavity)
3. Enlargement of the cavity in the area along the hollow space
4. Enlarging the hollow space by blowing it into the enlarged area

Two GITBlow method variants are distinguished: the one-step and two-step methods. In the one-step method, the cavity enlargement is achieved by a core pull. In the two-step method, the component is transferred into a larger cavity before blowing takes place.

3.10.4.1 One-Step Method

The process sequence of the one-step GITBlow process is depicted in Figure 3.37 as follows:

1. Injection
2. First injection of gas
3. Enlargement of the cavity by a core pull
4. Second injection of gas and cooling
5. Opening of the tool
6. Removal of the molded part

In Figure 3.38, an exemplary component made with the one-step GITBlow method is shown.

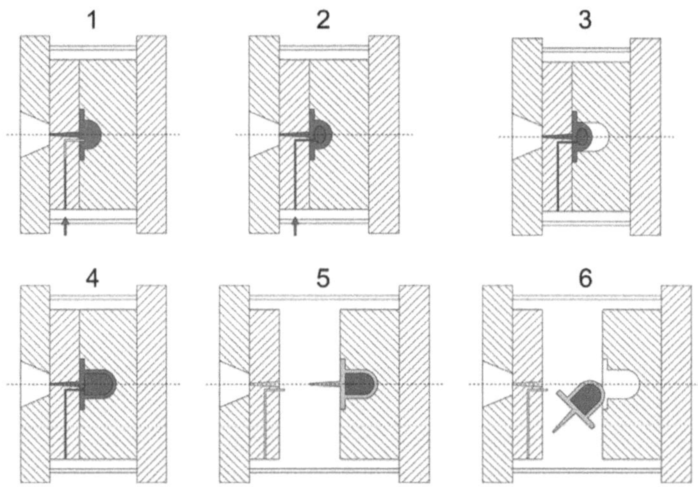

Figure 3.37 Schematic illustration of the one-step GITBlow process.

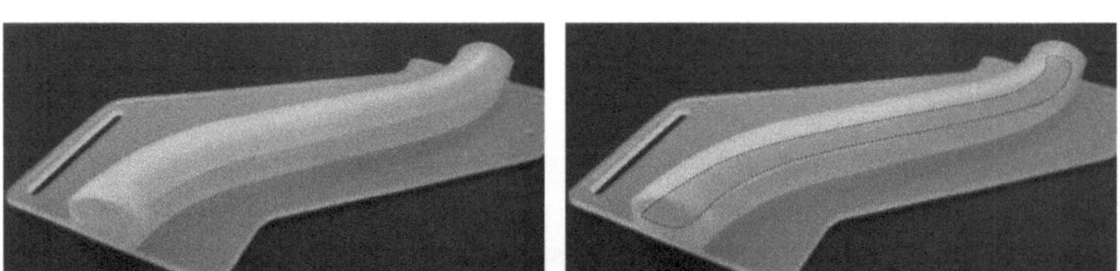

Figure 3.38 Exemplary component for the one-step GITBlow method; right: the original GIT hollow space has been marked.

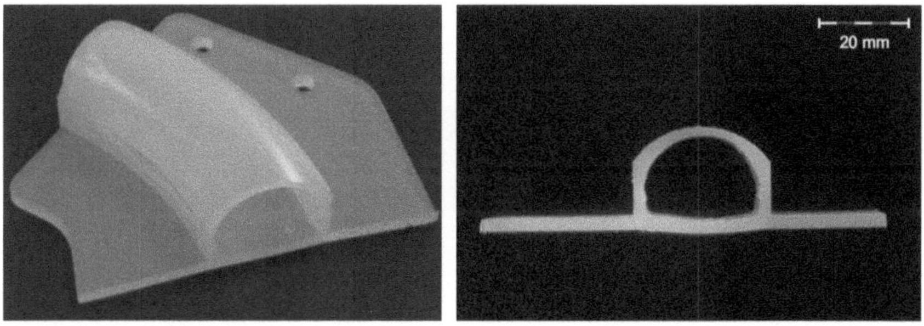

Figure 3.39 Depiction of a cross section of a component manufactured using the one-step GITBlow method.

In Figure 3.39, a cross section of a component manufactured using the one-step GITBlow method is depicted. The vertical flank, which forms due to the core pull, is clearly visible.

The one-step method has the following advantages:

- Relatively easy process control
- Lower tool costs

However, the following disadvantages and limitations result:

- Limited final geometry, because the preform can be stretched only in the direction of the movement of the core
- Straight, stretched flanks result for the blown geometry; that is, it is not possible to manufacture circular cross sections

3.10.4.2 Two-Step Method

The process sequence of the two-step GITBlow process is depicted in Figure 3.40 as follows:

1. Injection into cavity 1 and injection of gas
2. Opening of the tool and transfer into the second cavity
3. Possible reheating (required only in exceptional cases)
4. Closing of the tool
5. Injection into cavity 1 and injection of gas in cavity 2
6. Opening of the tool and removal from cavity 2

In the two-step method, two identical cavities with a rotary disk on the sprue side are needed, and two fixed cavities with differing volumes on the locking side are needed.

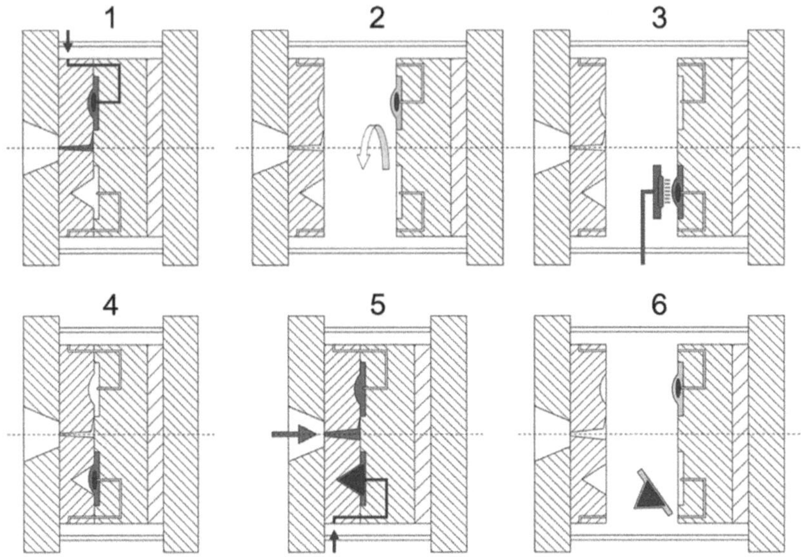

Figure 3.40 Schematic illustration of the two-step GITBlow process.

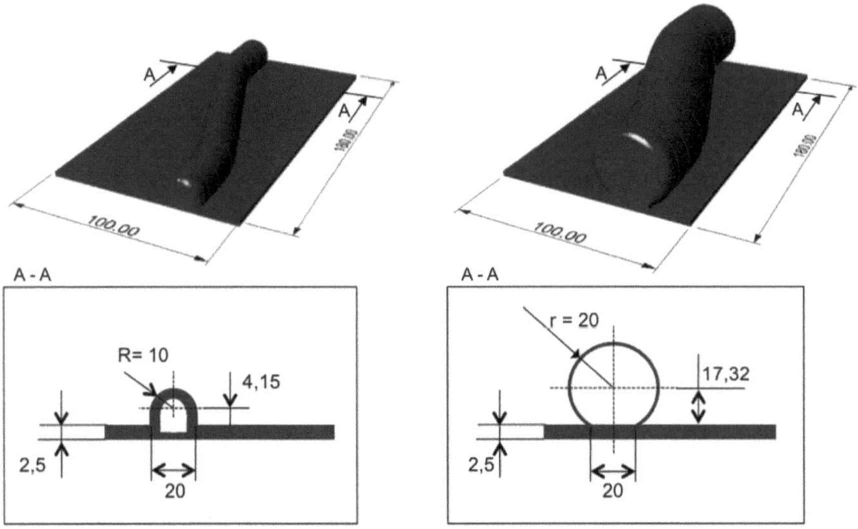

Figure 3.41 Exemplary component made using the two-step GITBlow process. Left: after the gas injection 1; right: after gas injection 2.

In Figure 3.41, an exemplary component manufactured using the two-step GITBlow method is depicted.

In Figure 3.42, a cross section of a component manufactured in the two-step GITBlow method is shown. The two-step method enables the

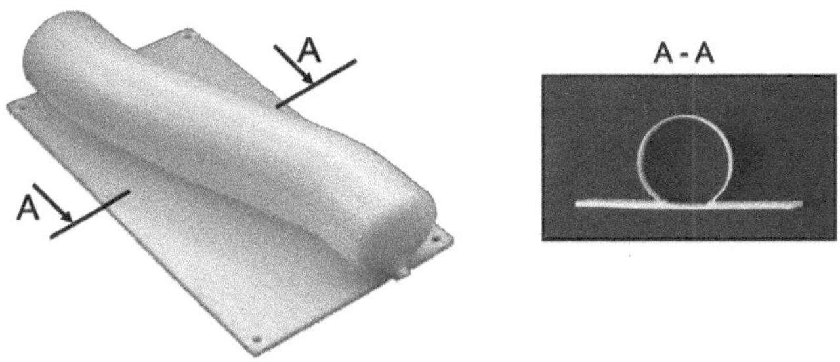

Figure 3.42 Illustration of a cross section of a component manufactured using the two-step GITBlow method.

creation of extremely thin-walled components, and the cross section of the final geometry can be completely different than the original geometry.

The two-step method has the following advantages:

- There is substantially more design freedom and fewer geometric limitations.
- The blown form can differ strongly from the preform.

The disadvantages/limitations of this method are:

- More complex process/process control
- Higher tool costs than for GIT

3.10.5 Multilayered Hollow Parts Manufactured Using GIT and WIT

When several plastic types, colors, etc. are to be combined with one another (as is possible when using sandwich technology), GIT or WIT can be used with the sandwich method. For this purpose, a skin component is injected first, then a core component, and finally gas or water, respectively. Multilayer hollow parts made using GIT or WIT are especially suitable for components in which the plastic cannot fulfill the requirements for the component by itself—i.e., combinations consisting of:

- Skin material with good mechanical properties, and core material with good chemical resistance
- Skin material with good mechanical properties, and core material with barrier properties

- Skin material with special haptic properties, and core material with good mechanical properties
- Hydrophilic skin material and water-resistant core material as a barrier in WIT

All variations of GIT and WIT can be used (partial filling method, side cavity method, etc.). The process control differs with GIT and WIT variants in terms of an additional step, in which the core component is injected. Process sequence, variants, the influence on the RWT, etc. have all been extensively examined and described in detail [30]. The process sequence is depicted schematically in Figure 3.43.

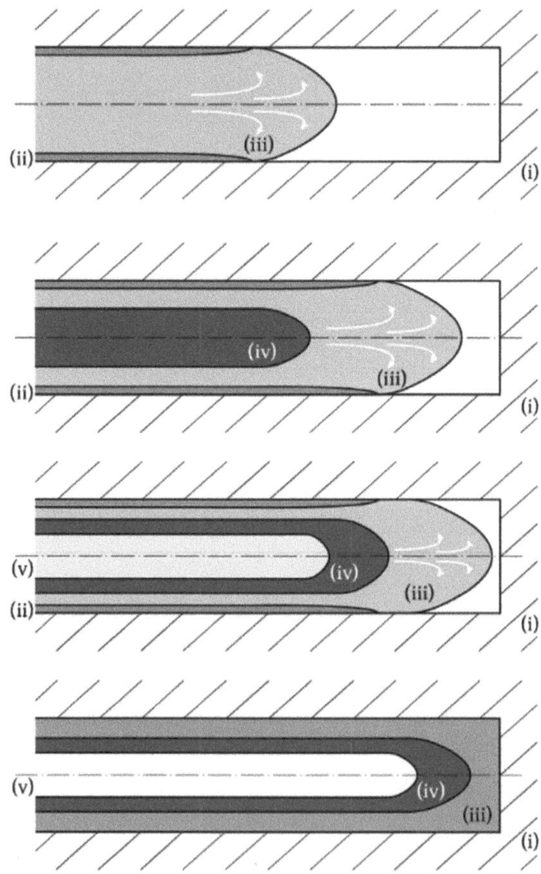

Figure 3.43 Schematic process sequence of a combination of sandwich injection molding and GIT or WIT. (i) Mold wall, (ii) frozen marginal layer, (iii) skin material, (iv) core material, (v) gas canal.

3.10.6 Measures for the Optimization of the Surface

Localized shrinkage compensation using gas, as well as the back-pressure method and the gas-counter-pressure method, are also on the list of GIT variants. However, they differ with regard to the process-technological principle.

In all cases, a compensation for the volume contraction during the cooling of the melt takes place. In the case of the local shrinkage compensation using gas, tool injectors are installed in the areas of the molded part at the greatest risk of shrinkage (thick-walled, or not reachable with a conventional holding pressure). Gas is injected to help preserve the hollow space inside the part and causes the surface to take the exact contour as the tool. Thus, sink marks, shrinking, and strong warpage can be avoided locally [54].

The back-pressure-method employs the same basic concept. However, a hollow space is not created. Instead, gas is injected between the already injected melt and the tool wall. Therefore, the molded part must fulfill the prerequisite for the application of this method, namely it must have a visible side, and a side that is not visible later on in the application. Gas is applied to this side and shrinkage is supported to deliberately create sink marks on the side that is not visible. The aim is to obtain a high-quality, visible surface [54,55].

By the gas counter-pressure method, the user can achieve an improvement of the surface of the injection molded part, and can avoid the changeover marks typical of the standard method, which are created due to short-term stagnation of the melt after the partial filling of the cavity. In this method, gas pressure is built up in the tool cavity to counteract the filling with melt. An even melt front results, as well as a controlled filling of the tool.

References

[1] Hobson JR. Method of and machine for forming hollow articles of plastic materials, United States Patent, US2331687 A, 12.10.1943.

[2] Friedrich E. Method for injection molding of hollow shaped bodies from thermoplastic resins, United States Patent 4 101 617, 17.08.1978.

[3] Mohrbach E. Verfahren und Vorrichtung zum Spritzgießen von Gegenständen aus Kunststoff, DE 2106546, 17.08.1972.

[4] Shah S. Gas injection molding: Current practices, ANTEC '91, pp. 1494–506.

[5] Rennefeld C. Konstruktive Optimierung von Thermoplastformteilen und Spritzgießwerkzeugen für die Gasinnendrucktechnik, Dissertation Universität-GH Paderborn, Shaker, Aachen, 1996.

[6] Moritzer E. Phänomenorientierte Prozeß- und Formteiloptimierung von thermoplastischen Gasinjektions-(GIT)-Spritzgießartikeln, Dissertation Universität-GH Paderborn, Shaker, Aachen, 1997.

[7] Johannaber F, Konejung K, Plaetschke R. Zufall oder Vorsehung? Ausbildung der Wanddicke bei der Gas-Injektions-Technik. Kunststoffe 1995;85(6):763–70.

[8] Märtins R, Diener L. GIT, 14. Stuttgarter Kunststoff-Kolloquium, Institut für Kunststofftechnologie und Institut für Kunststoffprüfung und Kunststoffkunde (1995) F4, S. 1–12, Stuttgart 1995.

[9] Eckardt H. Gas-assisted injection molding. In: Stevenson JF, editor. Innovation in polymer processing: molding. Munich, Vienna, New York: Carl Hanser Verlag; 1996.

[10] Gall M. Aspekte für Anwendungsmöglichkeiten von Gasinnendruck-Formteilen, Ludwigshafen, Seminarvortrag Seminar "Gasinnendruckverfahren beim Spritzgießen", Süddeutsches Kunststoff Zentrum, Würzburg, 1993.

[11] Steinbichler G, Knauder E, Langecker GR, Marsche M, Lake TG. Simulations programm, Experimentelle Überprüfung. Kunststoffe 1995;85(11):1894–8.

[12] Hansen M. Verfahrenstechnische Grundlagen zur Auslegung von Gasinnendruckformteilen, Dissertation Universität-GH Paderborn, Shaker, Aachen, 1996.

[13] Heim H-P. Qualitätsorientierte Betrachtung der Gasinjektionstechnik beim Spritzgießen, Dissertation Universität Paderborn, Shaker, Aachen, 2001.

[14] Findeisen H. Strömungsvorgänge im Werkzeug und der Einfluss der Materialeigenschaften bei der Gasinjektionstechnik, 17. Kunststofftechnisches Kolloquium des IKV, Institut für Kunststoffverarbeitung, F7, S. 65–70, RWTH Aachen, 1994.

[15] Michaeli W, Lanvers A. Spritzgießen transparent gemacht—Neue Entwicklungen bei der Prozeßsimulation, Teil 1: CAE-Techniken für das Zweikomponentenspritzgießen und das Gasinjektionsverfahren. Plaste und Kautschuk 1992;39(7):241–8.

[16] Sauer R. Experimentelle Erkenntnisse zur Gasinjektionstechnik (GIT), Seminarvortrag Seminar "Gasinnendruckverfahren beim Spritzgeißen", Süddeutsches Kunststoff Zentrum, Würzburg, 1993.

[17] Schröder T. Neue Aspekte bei der Herstellung von Kunststofformteilen mit der GIT, Dissertation RWTH Aachen, Verlag Mainz, Aachen, 1997.

[18] Turng L-S. Computer-aided engineering for the gas-assisted injection molding process. In: Stevenson JF, editor. Innovation in polymer processing: molding. Munich, Vienna, New York: Carl Hanser Verlag; 1996.

[19] Potente H, Heim H-P, Ridder H. The cross-section is crucial. Kunststoffe Plast Europe 2003;10:20–3.

[20] Steinbichler G. Gasinnendruckverfahren. In: Neue Werkstoffe und Verfahren beim Spritzgießen, Hrsg.: Verein Deutscher Ingenieure, Düsseldorf: VDI-Verlag Kunststofftechnik; 1990.

[21] Utescheny R. Gas-Injektions-Technik Ein Sonderverfahren als Standardlösung, SKZ-Fachtagung F10, Innovative Spritzgießtechnologien—ein Beitrag zur Konjunkturbelebung, Würzburg, 1993.

[22] Johannaber F. Gas-Injektions-Technik beim Spritzgießen (GIT); SKZ-Fachtagung, F9, Innovative Spritzgießtechnologien, ein Beitrag zur Konjunkturbelebung, Würzburg, 1993.

[23] Jaroschek C. Elegant? Preiswert? Oder sogar beides? Kunststoffe 1997;87 (9):1172–6.

[24] Potente H, Hansen M, Burgdorf D. Das GID—Verfahren am Beispiel eines Haltegriffes. Plastverarbeiter 1995;46(3):40–51.

[25] Schröder T. Gasinjektionstechnik für optimale Formteilqualität. Kunststoffberater 1995;85(9):39–43.

[26] Jaroschek C. Gasinnendruck zum Ausblasen überschüssiger Schmelze. Kunststoffe 1990;80(8):873–6.

[27] Eyerer P, Märtins R, Bürkle E. Spritzgießen mit Gasinnendruck. Kunststoffe 1993;83(7):505–17.

[28] Jaroschek, C, Utescheny, R: Möglichkeiten am Beispiel eines Pkw-Außenspiegels, Plastverarbeiter 42. Jahrgang 1991;42(5):55–7.

[29] Lanvers AP. Analyse und Simulation der Kunststoff-Formteilbildungsprozesse bei der Gasinjektionstechnik (GIT). RWTH Aachen, Dissertation, 1993.

[30] Lettowsky C. Herstellung zweischichtiger Kunststoff-Hohlkörper durch Gas- und Wasserinjektionstechnik, RWTH Aachen, Dissertation, 2009.

[31] Carreau PJ. Rheological equations from molecular network theories [Ph. D. thesis]. Madison: University of Wisconsin, 1968.

[32] Findeisen H. Ausbildung der Restwanddicke und Prozeßsimulation bei der Gasinjektionstechnik, Dissertation, RWTH Aachen, 1997.

[33] Renger M. Das Gasinnendruckverfahren—eine Spritzgießvariante mit besonderen Möglichkeiten, Süddeutsches Kunststoff-Zentrum Würzburg, Fachtagung 18–19, September 1990, pp. 101–36.

[34] Klotz B. Voraussetzungen im Bereich der Formteilgestaltung für die Anwendung des Gasinnendruckverfahrens, Transferzentrum Aachen Kunststofftechnik, Spritzgießtechnisches Kolloquium 1990, S. 36–57.

[35] Ridder H. Durchgängiges Konzept für die Auslegung von Spritzgießsonderverfahren am Beispiel der Gasinjektionstechnik, Dissertation Universität Paderborn, 2005.

[36] Potente H, Heim H-P, Ridder H. Accurate determination of the injection points. Kunststoffe Plast Europe 2003;7:21–5.

[37] Michaeli W, Brunswick A, Gruber M. Gas geben mit Wasser—Wasserinjektionstechnik (WIT): Eine Alternative zur Gasinjektionstechnik? Kunststoffte 1999;89(4):84–9.

[38] Michaeli W, Brunswick A, Pohl TC. Gas oder Wasser? Spritzgießen von Hohlkörpern durch Fluidinjektion. Kunststoffe 1999;89(9): 56–62.

[39] Michaeli W, Brunswick A, Kujat C. Kühlzeit reduzieren mit Wasser-Injektionstechnik—Vorteile gegenüber der Gasinjektionstechnik. Kunststoffe 2000;90(8):67–72.

[40] Michaeli W, Brunswick A, Pfannschmidt O. Gewusst wie! Fragen und Antworten zur Wasserinjektionstechnik. Kunststoffe 2002;92(9):94–8.

[41] Protte R, Konejung K. Wasser erhöht die Produktivität, Wasserinjektionstechnik. Kunststoffe 2003;93(2):52–7.

[42] Michaeli C, Jüntgen T. Injektortechnik als Schlüsselrolle—Werkzeuginjektoren für die Wasserinjektionstechnik. Plastverarbeiter 2002;53(10):152–3.

[43] Michaeli C, Jüntgen T, Brunswick A. Die WIT auf dem Weg zur Serie—Wasserinjektionstechnik jetzt industriell angewendet. Kunststoffe 2001;91 (3):104–6.

[44] Ridder H, Heim H-P. Verfahren zur Herstellung von Kunststoffformteilen, European Patent EP 1753592 B1.

[45] Heim H-P, Ridder H. GITBlow: Inflating Functional Channels in the Injection Mould; Kunststoffe Plast Europe, 95 (2005).

[46] Moore S. GIT derivative delivers precise channels. Modern Plastics Worldwide 2006;83:4.

[47] Potente H, Schöppner V, Schäfers M. Injection molding creates large hollow channels—the new, innovative gitblow technique. Society of Plastics Engineers, Annual Technical Conferences, Technical Papers, 2008.

[48] Potente H, Moritzer E, Schäfers M, Plugge T, Heim H-P, Ridder H. Advanced investigations for the newly developed special injection molding process: 'GITBlow', Society of Plastics Engineers, Annual Technical Conferences, Technical Papers, 2010.

[49] Ridder H. GITBlow: blow molding in the injection mold. Kunststoffe 2008;98:9.

[50] Lessmann J-S, Moritzer E, Plugge T. Description by dimensional analysis of the wall thickness distribution in the two-stage GITBlow process, ANTEC—Conference Proceedings, 2012.

[51] Moritzer E, Plugge T, Seidel S. Experimental verification of process-parameter-dependent temperature simulation of the two-stage GITBlow-process, Society of Plastics Engineers, Annual Technical Conferences, Technical Papers, 2013.

[52] Moritzer E, Plugge T, Seidel S. Contribution tot the development of a variothermal process control fort he two-stage GITBlow process based on the finite element method, ANTEC—Conference Proceedings, 2012.

[53] Plugge T. Herstellung komplexer polymerer Hohlkörperstrukturen. Ganzheitliche Prozessanalyse und -optimierung des Spritzgießsonderverfah-rens "GITBlow" Schriftenreihe Institut für Leichtbau mit Hybridsystemen 3, 2014.

[54] Stieler U. Universeller Problemlöser? Plastverarbeiter 1999;50(1):22–5.

[55] Eckardt H. Gasunterstütztes Spritzgießen—Einfallstellen bei dickwandigen Teilen vermeiden. Kunststoffe 1999;89(9):98–102.

4 Injection Molding of Multimaterial Systems

Tatjana Geminger and Stefan Jarka
Universität Kassel, Institut für Werkstofftechnik, Kunststofftechnik, Kassel, Germany

4.1 Introduction

The demand for lightweight construction and efficiency in many industrial sectors leads to the development of strategies (i.e., multimaterial designs) that utilize the property characteristics of the composite partners synergistically. Plastic–metal composites and organic sheet applications have great potential for both weight savings and enhancement of the products' mechanical properties. They not only provide a number of advantages in terms of weight and functionality in comparison to metal or plastic constructions, but can also result in more flexible, efficient production processes.

Depending on the desired application, various methods have been established, such as adhesive bonding, welding, assembly, coextrusion, molding, and insert molding. Injection molding technology was established many years ago in a variety of industries owing to its innovative design options and economical manufacturing process. In particular, hybrid designs made of plastic–metal sheet structures are often employed; the array of products ranges from compact electronic components to structural components of a large volume in the automotive and aerospace industries. Organic sheets can mainly be found in structural, automotive elements and sports equipment, but are increasingly employed in other areas as well. Composites with metal components that consist of a molded structure are not widespread. However, they are very important in household electrical appliances and automotive technology and are gaining in significance.

The strength of a composite is the key to its functionality. In the case of plastic–metal hybrids, it is essentially influenced by the process technology, process control, geometry of the molded part, and design of the bonding surface. In organic sheet processing, coordinated material selection and process control determine the quality.

Specialized Injection Molding Techniques.
DOI: http://dx.doi.org/10.1016/B978-0-323-34100-4.00004-3

This chapter covers two topics: plastic–metal hybrids and organic sheets. However, the individual sections on these topics overlap thematically. In the theoretical section, the basic mechanisms for the successful achievement of composite adhesion will be discussed. Understandably, the focus will be on the major points of the technology. Subsequently, a practice-oriented description of the established methods will be given.

4.2 Bonding Mechanisms in Plastic–Metal Systems

4.2.1 Adhesion

The term *adhesion* describes all possible interactions between two condensed phases, which induce a mechanical cohesion of the involved phases in the interface layer.

In the research literature, numerous adhesion theories have been introduced, but the studies not yet provided a complete description of the phenomena that occur in the interface. Rather, they should be regarded as an overlapping of various mechanisms. Essentially, a differentiation is made between specific and mechanical adhesion [1].

4.2.1.1 Specific Adhesion

The term *specific adhesion* describes the physical, chemical, and thermodynamic processes of the interfaces and entails six theoretical approaches (Figure 4.1):

- *Chemisorption* describes the development of various chemical bonds in the interface. In metal adhesions, chemical bonds were verified experimentally [2].

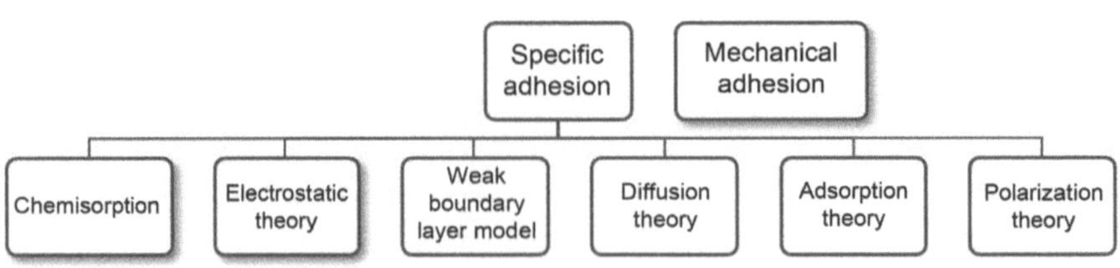

Figure 4.1 Overview of the adhesion theories.

- According to the *electrostatic theory*, an electric double layer develops between the joining partners, which promotes adhesive forces [2]. In plastic–metal composites, a relatively thin double layer develops, which promotes adhesion [3]. Additional studies concerning this approach have not yet been published.
- The *Weak Boundary Layer Model* describes the development of an intermediate layer between the composite partners made of impurities, which inhibits adhesion between them. This theory is often used if adhesion cannot be explained using other theories [4].
- The *diffusion theory* describes the interdiffusion processes between the two joining partners. In terms of plastic–metal connections, this theory is of minor significance [2].
- The *adsorption theory* is based on intermolecular and interatomic interactions in the interface; its prerequisites are a sufficient rapprochement and wetting of the phases [2].
- The *polarization theory* describes adhesion in the interface by means of the molecular–physical interactions. An adhesive bonding is only possible in the case of the same polarity [5].

The individual theories are elaborated in detail in Chapter 1 of this book.

The theoretical values for the physical and chemical intermolecular bonding forces are given in Table 4.1. The range of the physical bonding forces (at 3–5 Å) is higher than that of the chemical bonding forces at (1–2 Å). However, the intermolecular attraction forces are much stronger.

The values in this table refer to idealized, even surfaces, which do not exist in reality. In the case of two real phases that are in contact with one another, the bonding forces will occur only on the uneven sections of the surface that are directly in contact. Thus, numerous contact points are desired to achieve good adhesion; they can be created by increasing the temperature and pressure.

In plastic–metal connections, physical phenomena like hydrogen bonds and London dispersion forces are responsible for the bond. Due to the reaction with humidity, hydroxyl groups arise on the surface of the metal, which can form hydrogen bonds with polar plastics. In nonpolar plastics, London dispersion forces mainly contribute to the bond [8,9].

Before a plastic–metal compound results, the metal surface must be put into a state that promotes adhesion. Figure 4.2 depicts the schematic structure of such a metal's interface. The topmost layer—the adsorption layer—is an impurity, and it should be removed during pretreatment of

Table 4.1 Comparison of Bonding Energies and Ranges for Physical and Chemical Forces [1,6,7]

Type of interactions	Bonding energy [kJ/mol]	Range [10^{-10} m]	Theoretically calculated absorption forces [MPa]	Experimentally measured composite strength [MPa]
Physical bonding forces				
Dipole-induced dipole	≤2	3–5	35–300	15–25
Dipole–dipole	≤20		280–1750	
Dispersion forces	0.08–40.0		60–360	
Hydrogen bond	≤50		500	
Chemical bonding forces				
Ionic bonds	560–1000		30	15–25
Covalent bonds	60–680	1–2	17,500	
Metallic bonds	110–260			

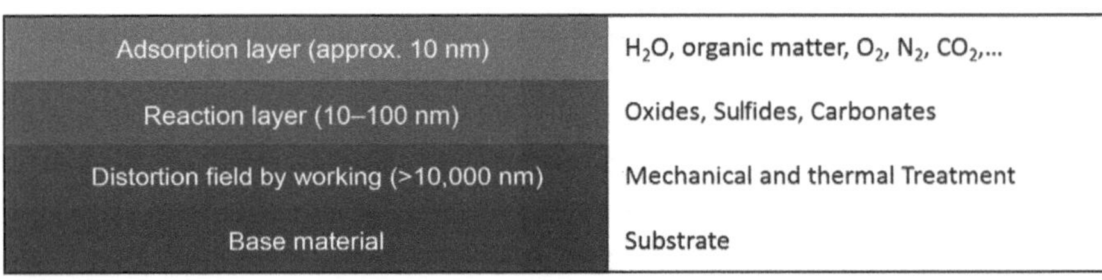

Figure 4.2 Schematic structure of a metal interface [10].

the metal surface. The reaction layer is a mixed layer consisting of metal oxides and metal hydroxides, which possess greater polarity than metals, and therefore bond with polymers [10].

Surface pretreatment for metals aims to remove impurities and adsorbed water. It is also used for oxidation, roughening the surface, and activating it [1] (Table 4.2).

Mainly, grease-dissolving substances are used to clean and degrease the surfaces. When using acidic cleaners, the surfaces are also freed of rust.

Mechanical pretreatment is intended for the removal of agents that inhibit adhesion and is used to roughen the surface. More information on this topic can be found in Section 4.2.1.2 of this chapter.

Table 4.2 Surface Pretreatments for Metals

Process	Treatment	Effect
Cleaning degreasing	• Organic solvents or alkaline cleaning agents • Spray degreasing • Steam degreasing • Ultrasonic cleaning • Immersion bath • Plasma process	• Cleaning of the composite partners
Mechanical pretreatment	• Grinding • Sandblasting[a] • Polishing	• Geometrical alteration of the surface → roughness • Removal of contamination layers
Chemical pretreatment	• Etching • Pickling with acid (concentration, temperature, time)	• Alteration of the chemical structure of the surface (oxidation)

[a]Sandblasting can increase the tendency for notch-sensitive joining components to develop stress cracks.

The metal surfaces are oxidized by chemical pretreatment, which creates an active layer with polar, functional groups that can form physical bonds (hydrogen bonds) [11]. If the joining process does not take place directly after pretreatment, a primer should be applied to the pretreated surface. The primer protects the surface from environmental influences, and functions as a connecting layer between the joining partners.

Generally, the effect or the selection of surface pretreatment cannot be transferred to all plastic–metal composites and must be individually adjusted to the composite partners.

4.2.1.2 Mechanical Adhesion

Mechanical adhesion represents a form-fitting connection that takes effect on a macroscopic level. The low-viscosity adhesive fills the pores, cracks, and indentations in the surface of a material, which results in mechanical anchoring between the two components once the adhesive has hardened [1] (Figure 4.3).

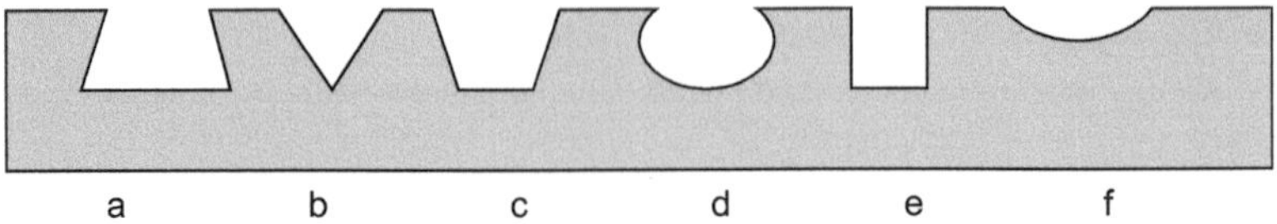

Figure 4.3 Schematic illustration of the geometries of roughness [11].

The penetration of the adhesive into the surface pores and the composite strength strongly depend on the geometry of the pores (meaning their depth, aperture angle, and diameter) [7]. Purely mechanical grouting occurs only in pores that have a type of indentation in the direction of the force, thus preventing deformation. This phenomenon is known as the *push-button effect.* Furthermore, the depth of penetration depends on the wetting properties of the adhesive, the surface tension, the viscosity, and the counterpressure of the gas enclosed in the pore.

Generally, a differentiation is made between micromechanical and macromechanical adhesion when classifying pores. Micromechanical adhesion involves pore sizes smaller than 0.1 μm [12].

Poggel was the first researcher who focused on the quantitative description of the composite strength in plastic–metal hybrids in correlation with the surface morphology [13]. A model for the quantification of indentations was created by means of mathematical, stochastic, and stereological approaches; it was placed in correlation with experimental strength measurements (traction-adhesive strength test), or finite element method (FEM) analyses. Thus, it was possible to demonstrate that indentation measurements are suitable for the characterization of mechanical adhesion in terms of strength. Moreover, it was shown that composite strength increases in positive correlation with an increase in the frequency and shape of indentations.

The composite properties of a metal surface, which was structured by a laser (DIN (Deutsches Institut für Normung) 1.4301) and had a molded-on plastic (PPS, glass fiber amount 40%) (Figure 4.4), were examined by Byskov-Nielsen in 2010 [14,15]. Using different sample geometries, the tensile and shear properties of the plastic–metal composite were determined in correlation with the pore distances and pore depths. The pore diameter remained consistent at 110 μm. In comparison to unstructured plastic–metal composites, the modified composites display a clear increase in strength. Moreover, a correlation between the pore distance or pore depth and the achievable strength was identified. However, the influence of specific adhesion was not taken into account in these studies, making these results only limitedly transferrable to other systems.

Hardly any quantitative models or methods exist for the description of mechanical adhesion in plastic–metal interfaces. Poggel [13] only

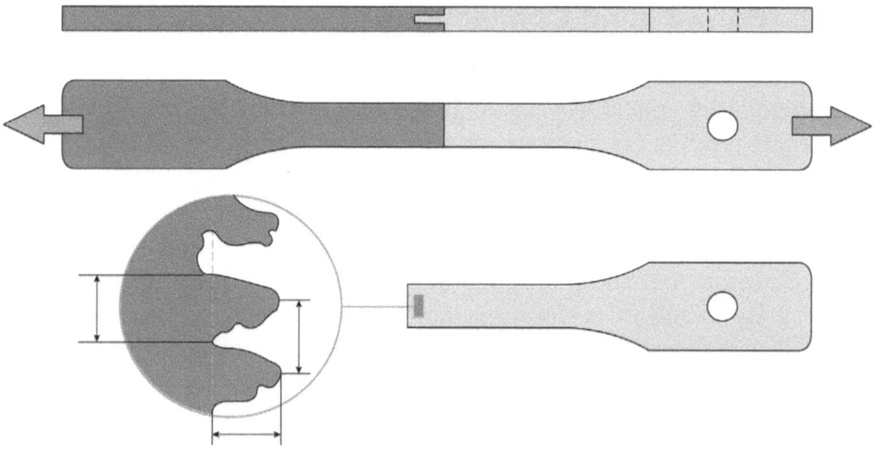

Figure 4.4 Metal surface that was structured by lasers and that has a molded-on plastic.

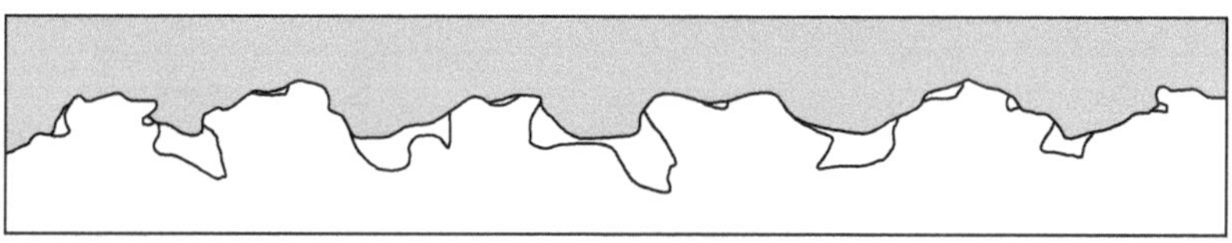

Figure 4.5 Remaining cavities in a rough surface [11].

dealt with the differences between specific and mechanical adhesion and their effects on the mechanical properties. Furthermore, the connection between the structure of the involved composite partners and the property changes of the materials are generally missing in the studies. In addition, the influence of the wetting properties of the adhesive, the surface tension, the viscosity, and the counterpressure of the gas enclosed in the pore have been insufficiently elaborated in observations until today (Figure 4.5).

In most cases, a surface enlargement of a surface intended for adhesion (i.e., by abrasive blasting, grinding, pickling, etc.) leads to an improvement in composite strength. The extent to which the mechanical grouting effect and the enlargement of the effective surface affect adhesive strength has not been fully clarified yet. It can be confirmed that the roughening of the surfaces promotes adhesion in most cases due to a high degree of surface activity caused by the creation of crystal lattice defects. Too much roughening can lead to decreased strength, which can be attributed to the development of stress peaks and the presence of cavities, and the reduced number of contact points as a result [7].

4.2.2 Form-Fitting Connections

Form-fitting and force-fitting connections represent another method of realizing plastic–metal hybrids in an injection molding process. In this process, these types of connections are achieved for plastic–metal components that can withstand high mechanical loads by using rivets, clinches, or beads.

The hybrid structural component called Erlanger Träger provides examples of various types of rivet and clinch connections between the metallic insert and the stiffening ribs made of reinforced thermoplasts [16,17]. By means of perforations and incisions made on the side of the metal insert, form-fitting rivet connections were created in the injection molding process. The preslitted sheets that bent open locally due to the melt pressure created clinch connections. In Figure 4.6, the preferential directions for the various types of loadings are shown. In contrast to rivet connections (cases a and b), clinch connections (cases c and d) can carry axial tensile, shear, and bending forces.

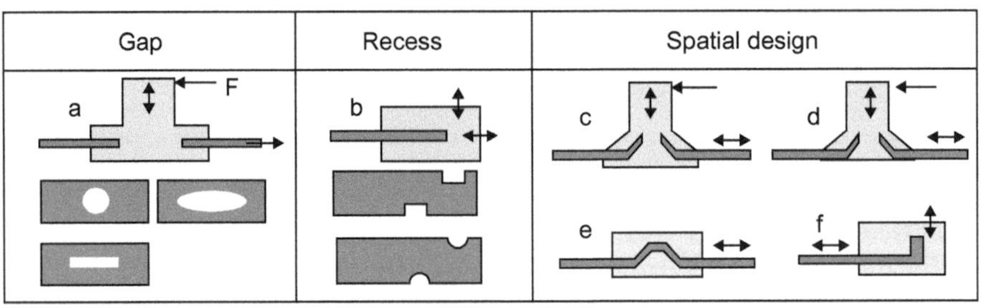

Figure 4.6 Connection options between plastic and metal and loading directions.

In contrast to rivet connections, bending the edges of the metal induces an additional grouting effect in the plastic component. This was verified experimentally in tensile-shear tests (Figure 4.7).

Another form of form-fitting connections that can be integrated into an injection molding process are beads [18]. These can be formed into the according forming cavity by means of the melt pressure (Figure 4.8).

The advantages of form-fitting connections over firmly bonded connections can be found in the area of connecting noncompatible materials. This is because a grouting effect takes place in the interface and a chemical reaction is not required. Another advantage of form-fitting connections is the high degree of automation in manufacturing. In addition, this type of connection makes it possible to transfer multiaxial stresses into the component. Firmly bonded connections (particularly adhesive bonds) are suitable only for the transferral of shear stresses.

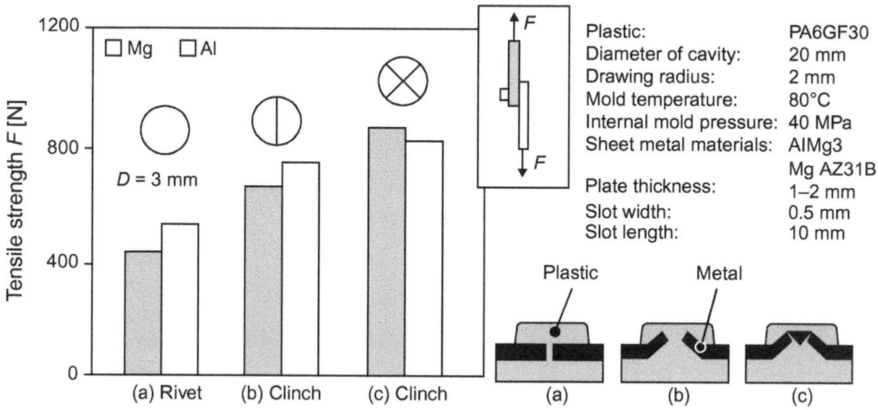

Figure 4.7 Tensile-shear strengths of rivet and clinch connections.

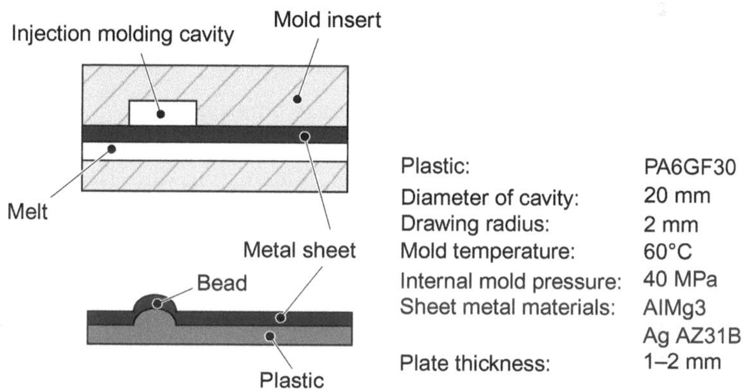

Figure 4.8 Creation of beads in an injection molding process with an integrated forming process.

4.3 The Influence of the Material, the Geometry, and the Process Parameter on the Component Properties

This section will focus on the most important aspects of the manufacture of plastic–metal hybrids in injection molding. Constructionally, there are two basic options to make a plastic–metal composite using injection molding. In the first option, the plastic component comes into contact with the metal component; the adhesion of the compound depends upon geometrical factors, like indentations and perforations, as well as on the shrinkage behavior of the plastic. In the second option, the plastic melt surrounds the metal insert, reducing the risk of

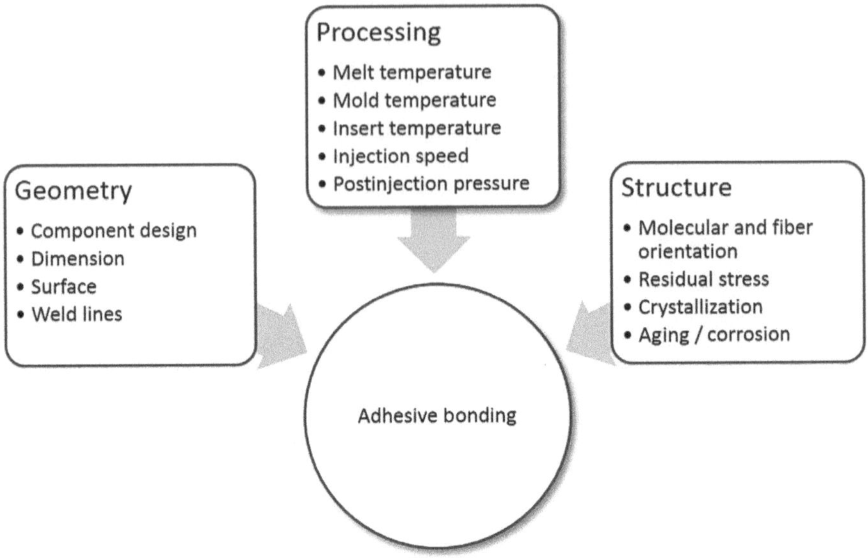

Figure 4.9 Parameters that influence composite adhesion.

detachment. However, other questions concerning the flow behavior, weld lines, and other elements must be considered. Furthermore, numerous interactions between the material properties and process parameters exist, which influence both types of hybrids. Thus, this text focuses on the essential material properties and mechanisms and their interactions, which are significant when manufacturing plastic–metal hybrids. A summary of these factors appears in Figure 4.9.

In the following sections, these factors and their interactions will be discussed with regard to the hybrid properties.

4.3.1 Construction

When designing and constructing hybrid components, material and construction criteria interact with each other so that the mechanical properties of the finished parts fulfill the requirements.

4.3.1.1 Selection of Materials

It is important to consider additional aspects when selecting a suitable plastic. This would be the environmental conditions of the finished part, special features of the injection molding process or the composite partners. In Figure 4.10, an overview of the influencing parameters is illustrated.

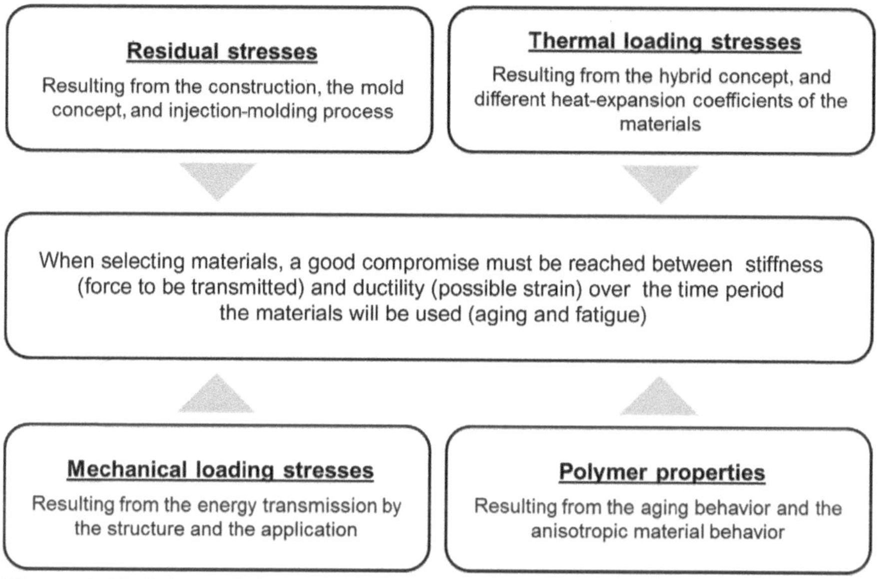

Figure 4.10 Schematic illustration of interactions for the selection of materials.

Internal stresses are an important issue for plastic–metal hybrids, and they are discussed separately in Section 4.3.2. In addition to the construction, the mold concept and the process influence, the selection of a material with a low thermal expansion coefficient already has a positive effect on the composite adhesion. In fiber-reinforced plastics, the beneficial orientation of the fibers can help reduce the difference between the thermal expansion coefficient of plastics and that of metals. Internal stresses always exist in a hybrid component; when media simultaneously affect the component (the effects of which differ depending on the type of plastics), stress cracks can form in the component. For this reason, semicrystalline plastics are favored for plastic–metal hybrids. In contrast to amorphous plastics, they display a much higher resistance to stress cracking.

4.3.1.2 Component Design

Despite the substitution of metal components by a plastic–metal composite, the overall stiffness of the component must be preserved

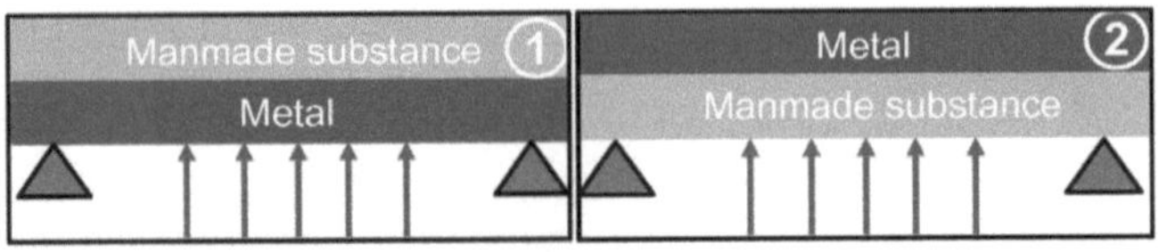

Figure 4.11 Two different options to absorb force in the sealing surface.

in accordance with the specifications. This especially applies to component areas in the sealing surface, because deformation would immediately result in leakage in the interface.

In plastic–metal composites, there are two ways to transmit force into the component. First, the forces can be intercepted by the metal component and get unloaded (Figure 4.11, #1, left). However, in this case, it is necessary to incorporate an intermediate layer in the sealing surface to prevent possible leakage. Second, the forces can be forwarded to the metal structure via the plastic component (Figure 4.11, #2, right). The mechanical loading capacity of the finished part depends on the metal structure and the arrangement and type of connection in the plastic component. The plastic component not only serves the purpose of integrating functions, but also acts as a reinforcing element in the form of ribs that prevent the metal sides from buckling or collapsing. The type of ribbing is individually designed to match the type of loading and the direction of loading.

Another aid when designing a component is a filling simulation for injection molding. In this way, component deformation that occurs due to form-filling behavior, the fiber orientation, the indentations, or the positions of the seams can already be identified in the construction phase. Uneven, asymmetrical filling of the form or uneven distribution of pressure can induce deformations in the metal insert.

Deformations of the metal component result in an uneven distribution of the wall thickness of the plastic components. Moreover, higher stresses remain in the metal insert after demolding. If these two effects overlap, substantial deviations from the original dimensions occur in the component. These issues can either be solved constructionally in the component or by optimizing the sprue system.

When the insert is surrounded by melt in the component, weld lines are inevitable. If possible, they should be relocated to areas that will not be subject to much loading because the joint line strength can be below 50% [19]. This large decrease in strength can be attributed to the stress concentration on the component surface, entrapped air in the weld line, or missing molecular entanglements.

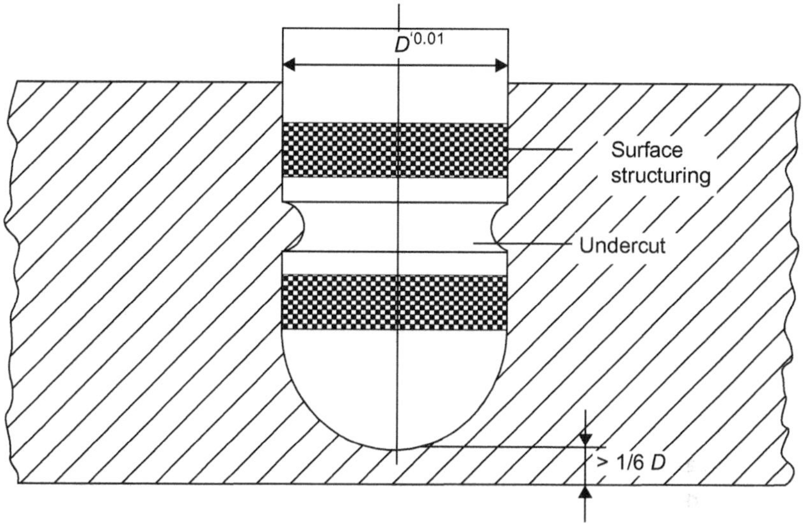

Figure 4.12 Design of a plastic–metal composite.

An important aspect of component design is the layout and pretreatment of the metal inserts. The formation of cracks is a major issue in plastic coating and can be attributed to internal stresses (Section 1.3.2). In order to keep internal stresses as low as possible, several guidelines should be taken into account during construction [20] (Figure 4.12):

- Avoid corners and edges on the metal inserts (no stress concentration).
- Equip inserts with indentations or structure the surface.
- Recommended allowances for the inserts are around ± 0.01 mm.
- The plastic thickness beneath the metal insert should equal at least 1/6 of the diameter of the metal insert.
- Employ impact-resistant, modified types of plastic.
- Set a sufficient wall thickness for the plastic coating s for large inserts (general rule: $0.3D \leq s \leq 0.4D$ for $D \leq 25$ mm).

4.3.2 Material Effects

4.3.2.1 Anisotropy

The mechanical properties of the components are strongly influenced by the existing orientations in the material. A differentiation is made between molecular orientations and orientations of the filling materials.

The molecular orientations are induced by the filling behavior in the cavity. The strong alignment along the mold and insert walls during the injection process and the subsequent, high cooling speeds cause freezing of a part of the oriented molecules. This leads to uneven shrinkage and anisotropic material behavior [19,21].

By this process, the orientation state of the material can be influenced to a certain degree, as follows:

- Increasing the melt temperature T_M promotes the relaxation capacity of the molecular chains, thus achieving lower orientation values.
- Higher temperatures in the mold also result in lower orientation values due to the same effect; however, it is less pronounced than with the melt temperature T_M.
- Higher injection speeds v_E lead to higher degrees of orientation in the injection phase. Faster cooling on the edge of the molded part causes them to freeze. However, the orientations relax faster in the center of the molded part owing to the shorter injection time and higher frictional heat. Consequently, in most cases, the orientation throughout the entire cross section of the component becomes smaller.
- An increase of the holding pressure leads to a greater degree of orientation. This can be explained by the fact that the orientations implemented in the holding pressure phase can barely relax due to the already starkly cooled melt.

For materials filled with fibers, the fiber amount and fiber orientation have a much greater influence than the molecular orientation. Figure 4.13 provides an example of the influence of the glass fiber orientation on the mechanical properties.

The tensile strength is higher in the direction of fibre orientation but visibly poorer than for the unfilled material when they are crosswise to the acting force. In particular, premature material failure can occur on the weld line due to the disadvantageous fiber orientation. Moreover, the orientations can cause uneven distribution of heat, and thus warpage of the component as well. For this reason, the fiber orientations should be adjusted to match the loading directions.

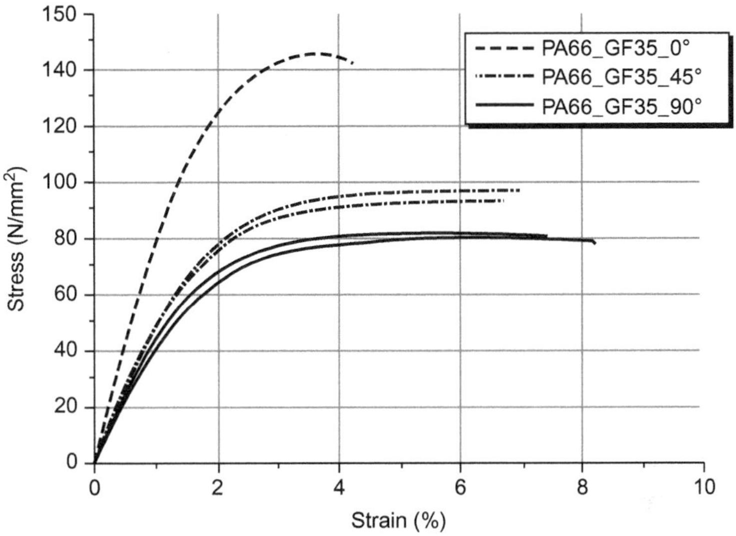

Figure 4.13 Influence of the glass fiber orientation on the tensile properties of a plastic sample.

4.3.2.2 Linear Thermal Expansion Coefficient

An important physical parameter that plays a role in the manufacture of plastic–metal hybrids is the linear thermal expansion coefficient α. It describes the length variation of the material depending on the temperature, and it also represents a function of the E-modulus. Plastics generally have a much larger thermal expansion coefficient than inorganic materials. When two joining partners have differing thermal expansion coefficients, the so-called $\Delta\alpha$ problem arises.

Figure 4.14 shows an example of the effect of the external, thermal load stress on the joining quality of the plastic–metal composite. The interfacial tensions arise, which eventually leads to adhesive fracture.

The thermal expansion coefficient of plastics can be reduced by specifically adding filling materials (particularly fibers), and, thus, can be matched to the thermal expansion coefficient of metals. It is important to note that the linear thermal expansion coefficient is smaller in the direction of fibre orientation than it is perpendicular to it.

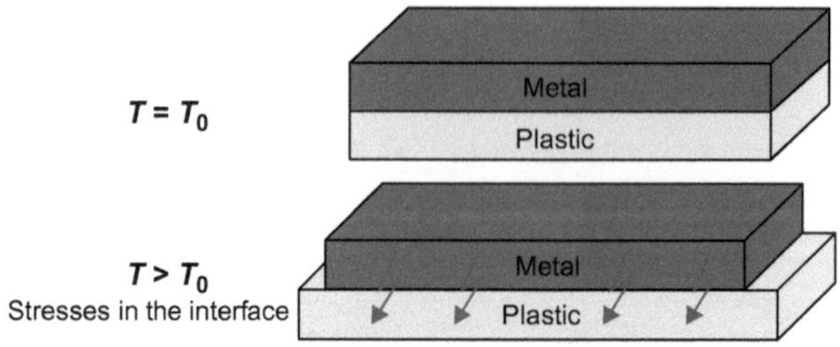

Material	Coefficient of thermal expansion α in 10^4 IK
Cast iron	9–10
Steel, unalloyed	11–12
Steel, alloyed	9–19
Copper	16–17
Aluminium	23–24
Thermoset	10–80
Thermoplastic	70–250
Glass	3.5–10

Figure 4.14 The effect of differing thermal expansion coefficients on composite strength.

4.3.2.3 Internal Stresses

Internal stresses are mechanical stresses that are present in the component without external influences. In injection molded thermoplastic components, internal stresses of the first grade primarily occur. This type of stress develops in the macroscopic areas and includes cooling/heating, holding pressure, and embedding stresses.

Cooling/heating internal stresses are present in all injection molded plastic components due to processing and are created by contraction of volume during cooling of the component (this effect is more pronounced in semicrystalline materials than in amorphous materials due to the crystallization process) [19]. The correlation between the pressure, the specific volume, and the temperature is referred to as *pvT-behavior*, and is illustrated in pvT-diagrams (such as Figure 4.15).

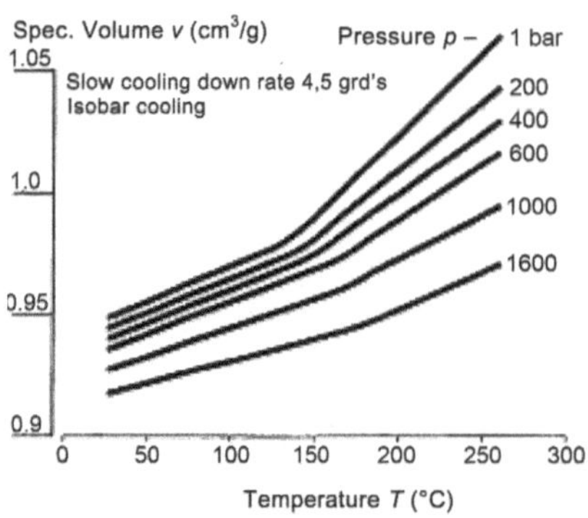

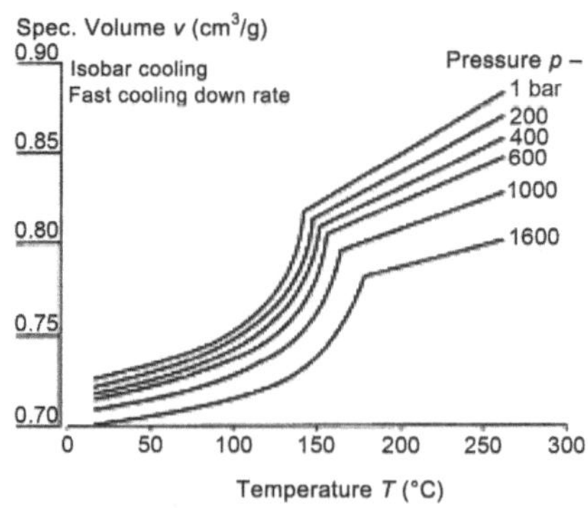

Figure 4.15 A pvT-diagram for the amorphous terpolymer ABS (left) and the semicrystalline thermoplast POM (right) [22].

In semicrystalline thermoplastics, fast cooling hinders crystallization in the outer layers. Inside the component, enhanced crystallization processes

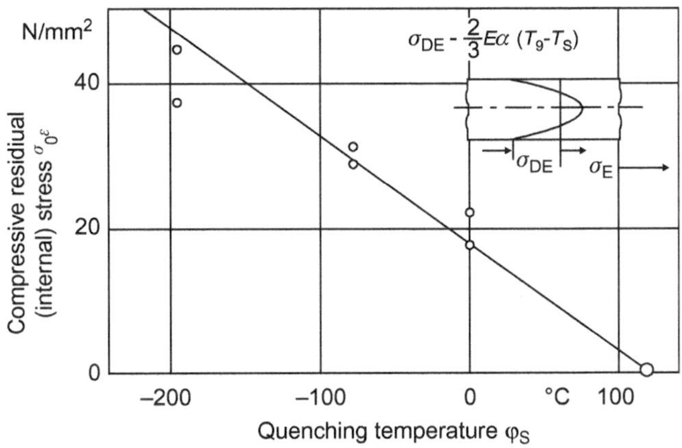

Figure 4.16 Internal compressive stresses on the surface of chilled bars made of PMMA [23].

take place due to delayed cooling; this directly correlates with a density increase. The density increase simultaneously indicates a reduction of the volume in the inner zones. These shrinkage differences lead to the development of internal tensile stresses on the inside and internal compressive stresses in the marginal area of the component (Figure 4.16).

In addition, internal thermal stresses occur when the following occur:

- A mechanical obstruction of expansion or contraction takes place.
- Material combinations have differing thermal expansion coefficients.

Over larger temperature ranges, the occurring thermal stresses in plastic–metal composites can be determined as follows [24]:

$$\sigma_T = -\int_{T_1}^{T_2} E_K(T,t) \cdot (\alpha_K(T) - \alpha_M(T)) \cdot \mathrm{d}T,$$

where

σ_T: Thermal stresses caused by a temperature change from T_1 to T_2

$E_K(T, t)$: Temperature- and time-independent elasticity modulus of the plastic

$\alpha_{K,M}(T)$: Temperature-dependent thermal expansion coefficient

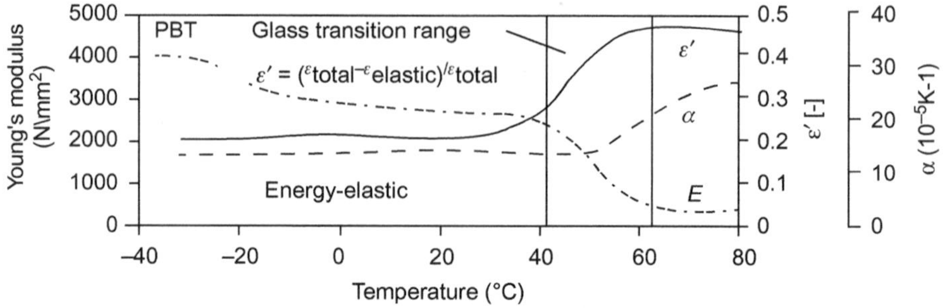

Figure 4.17 E-modulus, linear thermal expansion coefficient α and the relaxation capacity of plastic ε' in correlation with the temperature [17].

The relaxation capacity of plastics counteracts the formation of internal stresses. In order to illustrate this point, an example of the relaxation behavior of PBT in a wide temperature range is provided here. The relaxation behavior of the plastic ε' can be expressed by the relation of the nonelastic deformation ratio ($\varepsilon_{ges} - \varepsilon_{el}$) to the elastic deformation ratio $\varepsilon_{el.}$ This is shown in Figure 4.17. As the temperature increases, larger, temperature-dependent deformations occur in the material above the glass transition temperature. However, the developing stresses are relatively low due to the relaxation processes that initiate much faster (high relaxation capacity and low E-modulus). While cooling beneath the glass transition temperature, the E-modulus increases and the relaxation capacity decreases (Figure 4.17). Thus, the highest possible internal thermal stresses can be expected in the energy–elastic range (beneath the glass transition).

In general, it can be said that higher internal stresses are to be expected in the following cases [25]:

- A heterogeneous distribution of the temperature in the component.
- High cooling speeds.
- Inhomogeneous cooling conditions in the injection molding mold.
- A disadvantageously arranged holding pressure profile during the compensation of shrinkage.
- Hindrance of shrinkage due to the form.
- A high thermal expansion coefficient (or strongly differing thermal expansion coefficients of the composite partners).
- A high E-modulus.

In plastic–metal composites, great internal stresses occur in the component as a consequence of the cooling of the polymer in the injection phase. This is especially the case when the metal insert results in so-called form-bound shrinkage. High internal tensile stresses are the result. In addition, the relatively large difference in the thermal conductivity between metals and plastics causes an uneven cooling of the component in the mold, which in turn causes a significant increase in stress in the component volume due to $\Delta\alpha$.

4.3.3 Process Parameters

The mechanical properties of the finished plastic–metal parts largely depend upon process control or process parameters in the injection molding process. In this section, the essential process dimensions and their effects on the component and composite properties will be elaborated.

When insert-molding a metal insert with plastic, a contact temperature is set in the interface that is mainly influenced by the melt, mold, and contact temperatures. Further, an important criteria that must be taken into account are the contact conditions between the insert and the injection mold and the thermal material values. Because the last criteria depend more on the geometry of the finished part or are material-specific, they will not be discussed in the following sections.

4.3.3.1 Melt Temperature

The temperature of the melt and its chronological progress strongly influence the quality of the molded component and the adhesive strength between the metal insert and the plastic component. Belyi [26] showed that the adhesive strength between aluminum foil and PA6 significantly increases in correlation with a rising processing temperature. This phenomenon can be attributed to the improved wettability of the metal when the temperature of the plastic increases. The better the wettability, the higher the number of mechanical anchoring points on inhomogeneous parts of the metal surface.

Furthermore, higher melt temperatures can positively influence the joint line strengths [27]. General guidelines do not exist because the strengths are determined by the melt temperature differently depending on the material and the position of the joint line. However, Haufe [27] identified the median melt temperature profile as a parameter for the flow line strength. In [28] is demonstrated the influence of the melt temperature on the joint line strength by measuring the microhardness in the area of the joint line. The studies verified that the weld line can be localized precisely by means of the microhardness measurements. But evidence of smaller differences between the weld line zone and surrounding area was found at higher melt temperatures.

4.3.3.2 Temperature of the Insert

The preheating temperature of the metal insert strongly influences its connection with the plastic component, and thus the composite strength as well. This phenomenon was verified in hot press tests in which the contact temperature was significantly higher than during injection molding. The tests showed that adhesive strength strongly increases with increasing insert temperature because the viscosity rises in correlation with the temperature, causing the surface tension to reduce and wetting to improve [11]. In addition, preheating the metal insert reduces the discrepancy between the linear thermal expansion coefficient and the difference between the shrinkage of the plastic and the contraction of the metal. This positively affects the state of stress in the interface. In the insert-molding process, the time between the insertion of the metal part and injection of the melt should be as short as possible because the preheated insert loses its heat quickly due to the cool mold. Theoretically, if effects such as cooling shrinkage and stresses in the interface are disregarded, adhesive bonding between the components could be achieved. The contact temperature between the plastic and metal for an adhesive bond/composite should be above the softening temperature for amorphous plastics and the crystallization temperature for semicrystalline plastics [11,17].

4.3.3.3 Mold Temperature

The tool temperature significantly influences the cooling conditions of the molded part in the tool and the crystallization and shrinkage effects [19]. In semicrystalline plastics, an increase of the tool temperature has a rather disadvantageous effect. When the sealing point is reached, a great amount of heat is stored in the finished part, which immediately leads to greater shrinkage values. Thus, the risk of stress cracks is larger.

In contrast, the higher tool temperature can affect the joint line strength positively. The higher heat input could reduce, or even eliminate, surface notches that could lead to a failure of the component. Today, solutions exist for the selective heating of tool areas where the position of the joint lines was identified [29,30].

4.3.3.4 Injection Speed

The injection speed has a great impact on the degree of molecular orientation in the component and therefore influences the shrinkage and the development of a joint line. The strong molecular orientation results from fast cooling near the wall and insert. It is solidified in place, and it induces large differences in shrinkage between the component center and the component edge. Then again, the surface quality and joint line strength benefit from the high injection speed. When the metal insert is surrounded by hybrid components, joint lines inevitably result. At higher injection speeds, the risk of a notch and stress peaks developing on the surface is reduced. If the injection speed is set too low, poor welding of the melt flows results in the joint line, and the strength decreases. Yet, if the injection speed is too high, diesel effects (scorched air at the end of the flow path) can occur. It was investigated that the improvement of the joint line strength is neutralized by the diesel effect, or even impaired by it. Customized tool ventilation can help to remedy this.

4.3.3.5 Holding Pressure

The holding pressure time and level essentially influence the shrinkage effects in the molded part. Increasing the holding pressure considerably reduces the shrinkage of the plastic due to its elastic compressibility. Thus, the depth of possible indentations, which lower the strength of the entire component in form of surface notches, is reduced [17]. Moreover, the reduced shrinkage induces less stress in the plastic–metal interface (Figure 4.6), which, in turn, positively affects the composite strength.

In addition, a higher holding pressure makes it possible to weld the various flow fronts of the weld line together, and thus improve the surface quality and mechanical strengths of the finished part [19].

4.4 Processing Variants

At present, injection molding is the most relevant joining technology in the industry. According to Johannaber [19], overmolding of metal inserts differs schematically from other processes (Figure 4.18).

Insert molding technology encompasses three different process variants: insert technology, outsert technology, and plastic–metal hybrid injection molding. All three process variants share several guidelines that are process-relevant and suitable for the demands.

Figure 4.19 depicts the production chain of hybrid structures. The relevant steps entail the construction of the inserts, the positioning of the inserts in the injection molding tool, the insert-molding of the inserts, and the removal of the components. The relevant aspects or process parameters were discussed in Chapter 1 of this book.

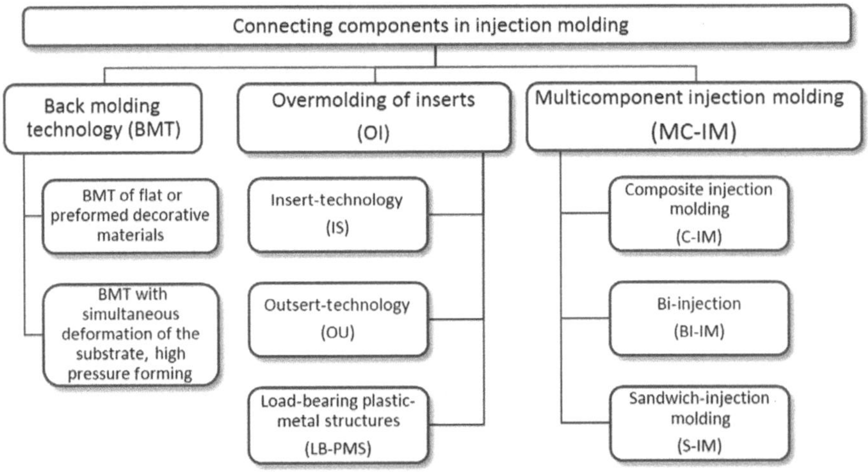

Figure 4.18 Classification of processes and the plastic–metal hybrid technology for the creation of connections in injection molding [19].

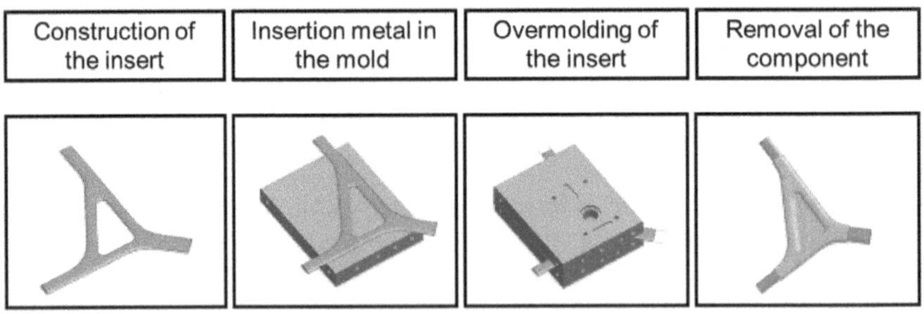

Figure 4.19 Process chain of insertion-molded metal inserts.

This section focuses on the various procedures and special features of the three process variants.

4.4.1 Insert Technology

In insert technology, various functional elements made of metal (insert nuts, screw nuts, plugs, wrench and screwdriver handles, etc.) are insert-molded with plastic in the injection molding process. Hereby the major part consists of plastic. It is in terms of its weight and volume and functions as a load-bearing structure for the metallic, functional parts (Figure 4.20).

In the injection molding process, the metallic inserts are placed into the recesses intended for them in the tool. Depending upon the design

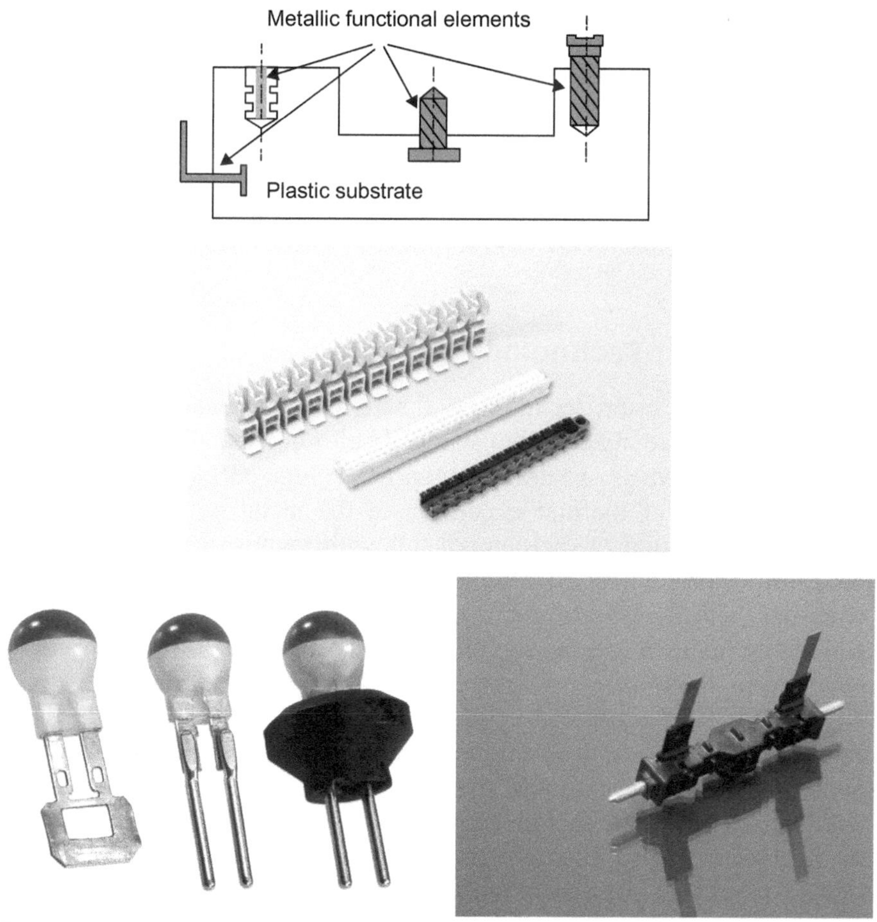

Figure 4.20 Plastic–metal connection in insert technology [56].

and position, they are fixed in the tool by sticks in holes, by pins, by support pins, snap devices, or magnets, or by suction using a vacuum. Due to the inserts, joint lines are inevitable in the finished component. However, their position can be calculated using the results from the flow simulation; this enables the joint lines to be positioned in a constructionally advantageous position.

When constructing the finished parts, the construction guidelines for metal inserts should generally be regarded. In order to ensure an inseparable connection between the plastic component and the metal insert, the inserts should have indentations, notches, dents, or perforations on flat pieces (Figure 4.21).

The insert hybrid components can be manufactured on conventional injection molding machines, but prolongation of the cycle time due to the insertion process should be expected. One possibility to decouple the injection molding process from the insertion process is provided in form of injection molding machines with integrated turning or sliding platforms. For example, tools with two halves on the lower side are employed. One half is intended for the insertion of the inserts. Further options for decoupling the two process are rotary disk and cube molds. Vertically constructed injection molding machines offer additional savings because the inserts are held in place by gravity [19].

4.4.2 Outsert Technology

In contrast to insert technology, functional and connecting elements made of plastic are injected into stamped holes in a metallic carrier board in a process known as *outsert technology*. The high stiffness, low density, and low degree of thermal expansion of the metal board increase the application potential of the components with regard to the dimensional behavior when subject to temperature loading. The reduction of the workload, which results from the installment of various elements in a single step in the injection process, affords cost savings for the individual components when using outsert technology. For a long time, outsert technology has had widespread application in many industrial sectors. For example, outserts are used in consumer electronics such as CD drives, household appliances, window regulators and sunroof mechanisms in the automotive field, the mobile communication market, and energy and medical technologies (Figure 4.22).

The employed materials must fulfill among others the following specific requirements. Cold-rolled steel, aluminum, or copper sheets are used for the metal boards. The thickness allowance of the sheets should

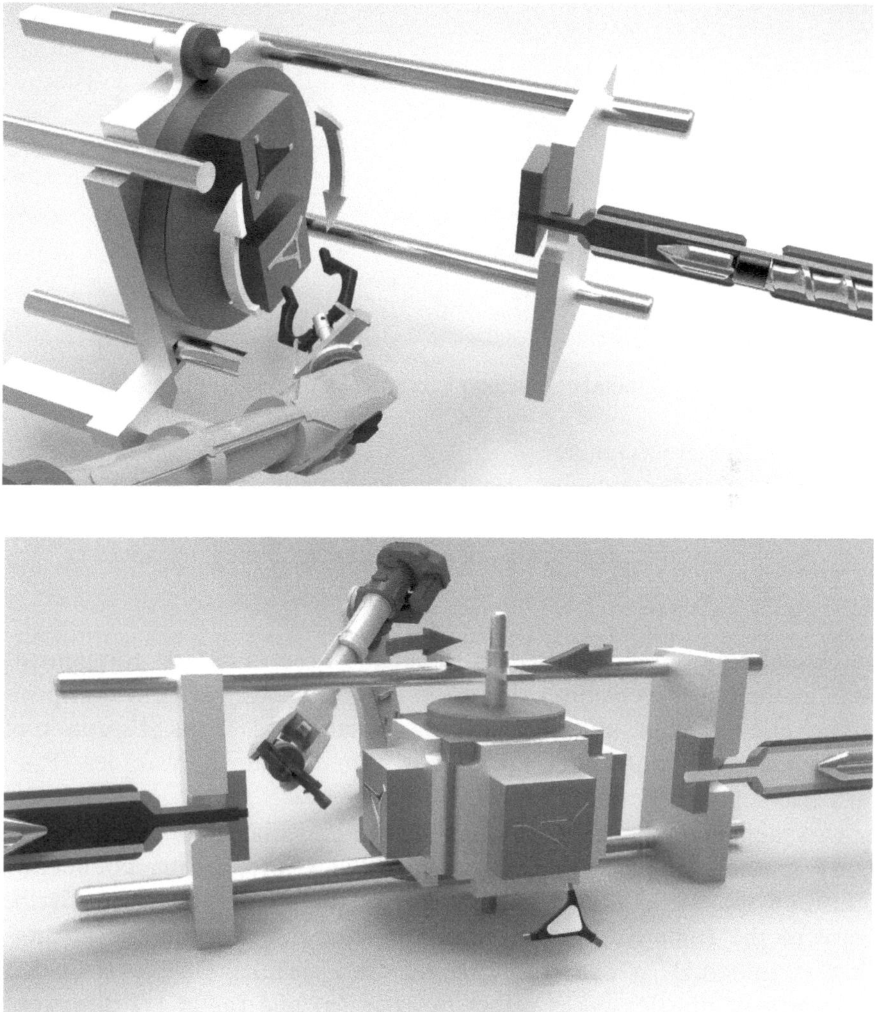

Figure 4.21 Rotary disk (above) and cube mold (below).

be no more than 4%; otherwise, problems will arise when closing the mold. Before the sheets are placed into the injection molding mold, they are deburred, cleaned, and, if necessary, preheated.

Depending on the area of application, plastics must fulfill different requirements, especially if multifaceted functional elements made of the same material are realized in one component. The employed plastics should have the following properties [17,19,31]:

- High strength, stiffness, and viscosity
- Good frictional behavior and a low frictional coefficient

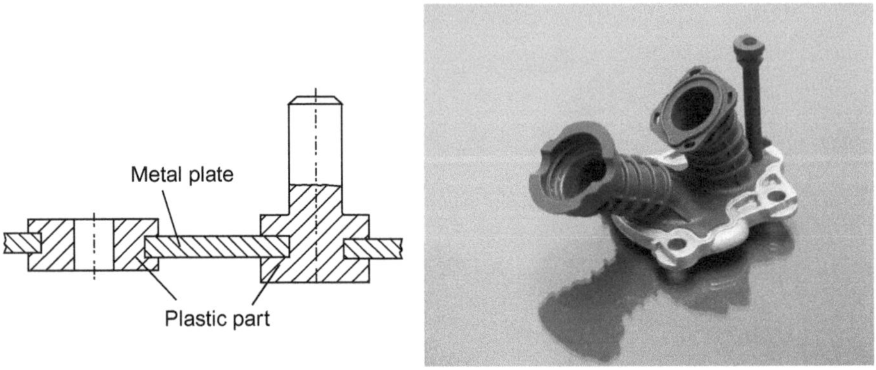

Figure 4.22 Plastic–metal connection in outsert technology [56].

- A low creep tendency.
- A high-dimensional stability when subject to changing temperatures and/or humidity.
- A high chemical resistance and low tendency to develop cracks.

Acetal copolymers (POM) are frequently used in outsert components due to their relatively high shrinkage (approximately 2%), which positively affects the grouting effect in the metal board. If the absorption of surrounding moisture does not play a primary role, polyamides (PAs) are also suitable for application.

An inexpensive, constructive execution is decisive for the functionality of the components. A form-fitting and force-fitting connection between the metal board and the plastic element is optimal. In the case of the force-fitting option, as explained next, a double anchoring is intended in order to prevent free shrinkage. The shrinking behavior of plastics takes effect in axial (a) and radial (r) directions (Figure 4.23, left). Shrinkage in the axial direction creates a force-fitting connection; in the radial direction, it leads to the formation of a gap. When subject to loading in this direction, the functional element shifts by the extent of the gap. This can be remedied by setting up several symmetrically organized connection points (Figure 4.23, right). In long, functional elements, stresses induced by the hindrance of shrinkage can lead to a deformation of the board. This can be solved constructively by dividing the functional elements into several, densely situated fragments, or by means of tensile cores (mold-technological).

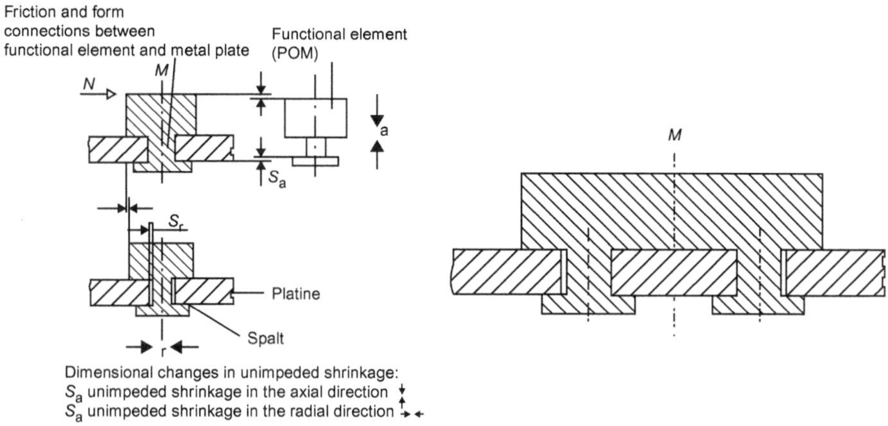

Figure 4.23 Shrinkage in a plastic element [31].

The following special features are also referred in the literature concerning the mold construction for outsert components:

- In order to avoid dimensional fluctuations in the component, the boards must always be uniformly aligned. This can be ensured by holding pins in the injection mold.
- Process-technologically, standard machines are used to manufacture components, but a mold with hot runner technology, or a rotating platform is required.

4.4.3 Load-Bearing Plastic–Metal Structures

Plastic–metal hybrid injection molding is a technology for the manufacture of large, lightweight structures with numerous design possibilities. Widespread application of this technology can be found in mechanical, structural components with a high loading capacity in the automotive sector. The material combination consists of formed metal sheets, which fulfill all mechanical requirements, and molded plastic components as well. The plastic component either fulfills the visual and design requirements or functions as a reinforcing element in the form of a molded, ribbed structure [17,18,32–35].

In contrast to the insert and outsert technologies, the plastic component is molded cohesively, which entails significant advantages with regard to the stiffness requirements. Moreover, unlike purely plastic solutions, the fitting accuracy of these hybrid components does not depend on influences such as material shrinkage and temperature fluctuations as strongly owing to the load-bearing role of the sheet metal structure.

Hybrid structural components are already applied in series production in the automotive sector; that is front end modules [16,36,37], mounting supports [38], hybrid doors, tailgate modules, roof frames, and brake pedals [37].

The loading capacity of the hybrid structure largely correlates with the material properties of the joining partners. According to Ehrenstein [16], the flexural rigidity and torsion stiffness of the finished parts increase as the stiffness and the wall thickness of the sheet metal components do. This was verified by means of bending tests on U-sections made of aluminum and zinc-coated steel sheet metal. In addition to a high level of stiffness, a higher elongation at yield would be desirable in the plastic component, so as to have a deformation reserve should overstressing occur. Moreover, the selected plastics should display a high resistance to stress cracking. Mainly, fiber-reinforced plastics are employed in hybrid structural components. The fiber reinforcement reduces the discrepancy between the thermal expansion coefficients of the materials (Figure 4.17). As a consequence, lower thermal internal stresses result in the component. The previously mentioned anisotropy of materials filled with fibers must also be taken into account.

The constructive connection design of ribbed structures complies with the stress profile of the finished parts. In Figure 4.24, examples of possible options to connect the plastic component to the metal profile are shown. They can be optimized to be suitable for demands using finite element method. The points where the ribs intersect simultaneously function as connection junctions to the sheet metal

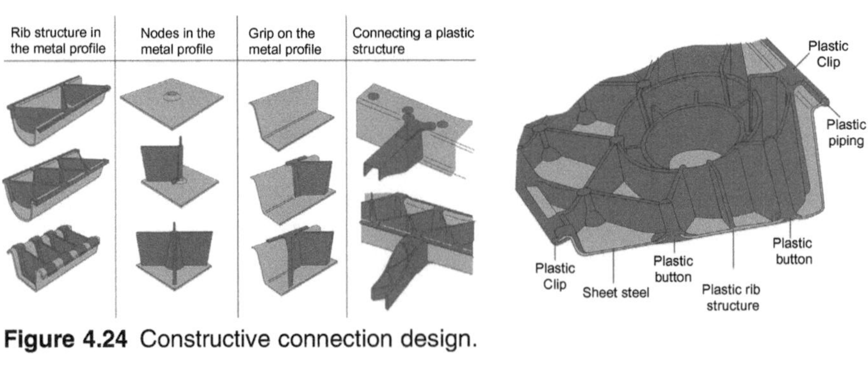

Figure 4.24 Constructive connection design.

profile. The ribbed structure is clinched to the sheet metal flange. Moreover, additional plastic elements can be attached all over the sheet metal structure.

In Section 4.2.4, further form-fitting connection options were illustrated. They are all applicable for hybrid technologies (insert, outsert, and hybrid injection molding). The direction of loading of the junctions is an important aspect to consider in the constructive design of hybrid connections.

Perforations and indentations in the metal component ensure a form-fitting material combination of the elements. An additional, form-fitting connection of components is achieved using a coupling agent, which enables a higher degree of adhesion between the components, as well as material savings of up to 20% [39] (Figure 4.25).

As was also the case for the previously mentioned methods, conventional injection molding machines are suitable for the manufacture of structural components in plastic–metal hybrid injection molding technology. The procedure is similar to the outsert process. A preformed, prestamped, and, if necessary, preheated sheet metal is placed into the injection molding mold and is overmolded with plastic. The metal inserts should be fixated in the mold so that they cannot be displaced during the injection process. Furthermore, it is recommended to use fault-tolerant positioning elements in the mold and on the insert to prevent dimensional changes, which occur due to the temperature differences during the injection process, from inhibiting the function of the component.

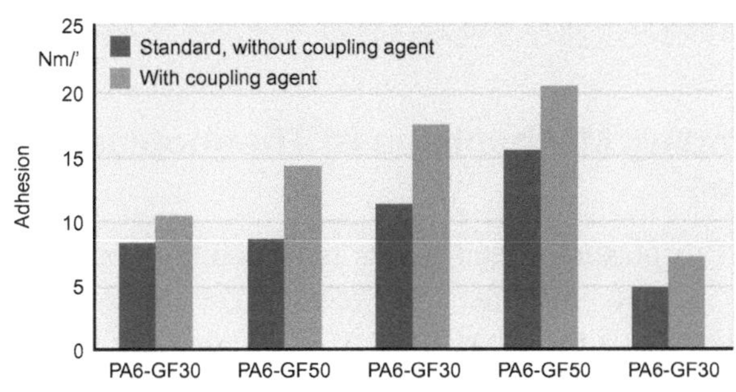

Figure 4.25 Profile cross section of a hybrid structure with special construction elements. Influence of the coupling agent on the adhesion in the conditioned "Erlanger Träger" [39].

4.5 Production of Injection Molded Organic Sheet Parts

Fiber composite technology, in combination with thermoplastic injection molding, is increasingly being employed in lightweight applications. Organic sheets are increasingly being used in this field. The benefits of the organic sheet processing options, which are suitable for large series, are mainly attributable to the thermoplastic matrix [40]. The material selection, the component design, and special manufacturing process must be coordinated with one another, if necessary, by means of simulation methods [41]. In sum, the following advantages in terms of weight and functionality in comparison to metal or plastic constructions can be named:

- A force- and form-fitting connection without a clear separating layer [42,43]
- Anisotropic behavior for application-specific reinforcement [43]
- A high mechanical loading capacity beyond its boundary layers (flexural rigidity, torsion, tensile strength along the orientation of the fibers) [37]
- Weight reduction due to low densities [44]
- Recycling is especially simple owing to a smaller variety of materials [42]
- The integration of functions in the injection molding process [40]

Figure 4.26 depicts a technology demonstrator, which is a typical, structural component application in vehicles.

4.5.1 Bonding Mechanisms of Thermoplastic Composites

In contrast to plastic–metal bonds in which the bonding mechanisms concentrate in one interface between different materials, a bond between materials of the same type is on hand when organic sheets are employed. Therefore, the selection of material must be planned, in order to be able to plan a functioning hybrid component. Ideally, the material and form-fit interact to create a connection that is not inferior to the strength of the basic matrix material.

Figure 4.26 Dynamically loadable structural component made of organic sheet for application in the automotive sector.

4.5.1.1 Types of Adhesion

Essentially, a very good connection can be expected when the same material is employed for the injection molded and matrix components. When using material variations, the melt temperature and the viscosity must be taken into account in addition to the compatibility of the materials [45]. The reflected-light microscopy image (Figure 4.27) shows the connection area of an injection molded component that is reinforced with an organic sheet. In the upper section, the short fibers of the injection molded component are discernible. In the lower right and left areas of the image, the oriented fibers of the organic sheet are visible. A clear separation of the source materials is not distinguishable, which underlines the achievement of a cohesive connection. The adhesion mechanisms between the thermoplastic materials (chemical mechanisms caused by primary valence bonds, physical mechanisms caused by secondary valence forces, and mechanical mechanisms caused by interlocking of the molecules) are described in detail in Chapter 1.

In addition to the three typical adhesion mechanisms of a welding connection, with which the cohesive connection can be compared locally (contact-induced adhesion, molecular stretching and relaxing, and self-diffusion), another phenomenon can occur. If the matrix is sufficiently heated and the plate thickness is rather low, perfusion processes can create a firmly bonded connection by penetrating the endless fibers of the thermoplastic composite insert. In Figure 4.27, it is evident that the injection molding component is located between the fiber strands of the organic sheet, and, thus, additionally creates a cohesive connection. Process steps that must be taken into account, like

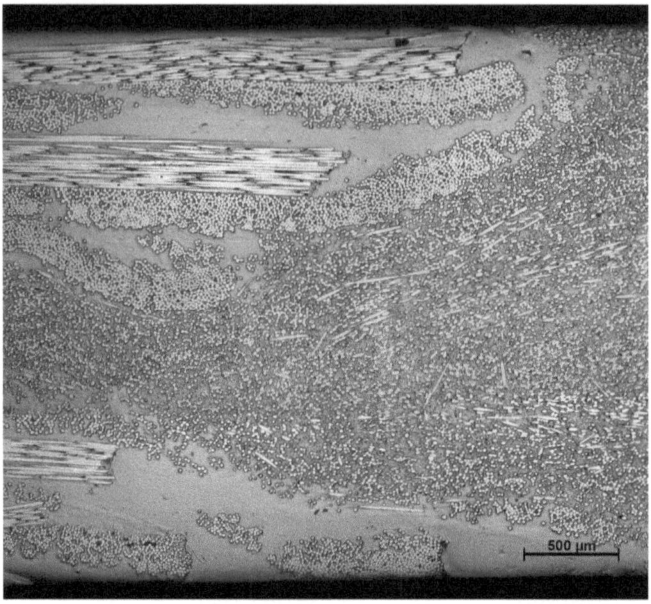

Figure 4.27 Microscopic image of the cross section of a thermoplastic composite between the organic sheet (bottom right and left) and an injection molded component reinforced with short fibers (top).

preheating the semifinished products, are explained in detail in the section entitled "Process Sequences," later in this chapter.

4.5.1.2 Bonding at Internal Stresses

Internal stresses and the warpage created as a result must always be taken into account during injection molding and can be reduced by means of process optimization (generally, by means of reducing the injection speed and increasing the melt and tool temperatures). These phenomena are often enhanced by employing differing materials, especially when adhesion, or geometrical indentation exist between them. This can be explained by the varying thermal expansion coefficients. Stresses are already generated during subjection to high temperatures in the manufacturing processes and during the subsequent cooling thereafter.

When utilizing organic sheets for hybrid components, these effects can be significantly reduced. Because the organic sheet can consist of the same fiber and matrix material as the injection molded component, at least the material-specific expansion coefficients are very similar.

However, a smaller length variation is on hand for the oriented endless fibers of the organic sheets than for injection molded types reinforced with short fibers. Unfilled granulate displays an even larger difference to the behavior of organic sheets. Table 4.3 shows selected materials and their expansion coefficient.

Table 4.3 Materials Relevant for Hybrid Components and Their Expansion Coefficients (Partially Exemplary Manufacturer Information)

Material (GF—glass fiber, GF 30—glass fiber ratio 30%)	Linear expansion coefficient α (10^{-6}/K)
Steel, alloyed	9–19
Aluminum	23–24
PA6	85
PA66	85
PP	135
PA6 GF 30	20–30
PA66 GF 30	25–35
PP GF 30	60
Organic sheet, PA6 and GF	15–19
Organic sheet, PA66 and GF	13
Organic sheet, PP and GF	15–19

Possible internal stresses in the component after subjecting the organic sheet to forming processes should be considered. Temperature control and the tailoring of the insert geometry strongly influence the manifestation of internal stresses [46].

Basically, organic sheets positively influence the warpage and dimensional stability of a component, and there are no real adhesion problems between the components when using the processing options. Because applications generally are in the field of structural components. Examinations of the leakproof behavior of the connection are not known yet.

4.5.2 Process Technology

Principally, very similar process technologies are utilized for the manufacture of organic sheet-reinforced components. Injection molding machines are the main production facilities, and individual stations are arranged around them. The process is characterized by a high degree of automation because thermoplastic components are always series with large quantities due to the high-investment costs for the production facilities.

Above all, the realization depends upon the component size, component design, and the quantity to be manufactured. The injection molding technological solution, which is generally regarded as the most complex portion of production, is realized individually for each application. Hereby, the tool for parts with very simple geometry can resemble a standard injection molding tool. More complex forms (complicated preforms, indentations, or connection points) that have to be directly molded, must be produced on complex tools with movable elements. In particular, forming the layer insert and positioning it in the injection molding tool require special solutions [47].

The size of the planned component is also a significant factor, because the projected area of the injection molded part directly correlates with the clamping force and the size of the machine in injection molding. Because organic sheet applications are usually structural elements, they often have an above-average size for injection molding conditions. Vertical injection molding machine technology, which is advantageous for better handling of the inserts, reaches its limits with larger-component applications. For this reason, it is also less common than conventional horizontal injection molding machine technology.

4.5.2.1 Process Structure

Due to the required machine technology, the organic sheet process can no longer simply be referred to as a special injection molding method. This is especially applicable for the preparation of organic sheets, since cutting, preheating, and preforming increase the complexity of the manufacturing process.

From a processing point of view, realization can be categorized into two variants. Here, the differences are essentially limited to the preforming process. With regard to the tool technology, upstream thermoforming is easier to realize. Another procedure includes forming the reinforcement layer in the injection molding tool [43]. Depending on the component geometry, special tool technologies are to be employed. Reasons for selecting a certain variant correlate with the component geometry, the cycle time, the quantity, and the necessary equipment investments. Moreover, there are various advantages; they will be described in the following section.

In terms of process control, the precisely repeatable tempering and positioning of the accurately cut organic sheet in the injection molding tool are of decisive importance.

4.5.2.1.1 Preforming Process

The conventional variant involves preforming, also referred to as *draping*, of the cut-and-heated organic sheet by means of an upstream thermoforming process. The geometry, which is inserted into the injection molding tool later, is created. The preforming process can be decoupled from other steps. In that case, a second heating to enable a load-bearing, firmly bonded connection between the sheet and injection molded component is unavoidable. Owing to the external preforming, the complexity of the injection molding tool can be minimized. Yet, the entire cycle time is longer than in the second variant. The process cycle is organized as follows:

- Creation of a organic sheet cut to size that correlates with the form of the component.
- Heating of the organic sheet.
- Preforming of the organic sheet.
- Preheating and insertion into the special injection molding tool.
- Injection molding of the plastic injection molding component.
- Removal of the component, which does not require post-processing in most cases.

Figure 4.28 shows the process cycle of the described variant.

Further processing of the semifinished product ensues similar to the molding of the insert, except for the required heating to ensure composite adhesion and the correlating, complicated handling of the semifinished product (see Section 4.5.2.2.2, later in this chapter).

4.5.2.1.2 Integrated Forming Process

Alternatively, preforming can also be carried out in a combined forming and injection molding tool. Consequently, heating must only be

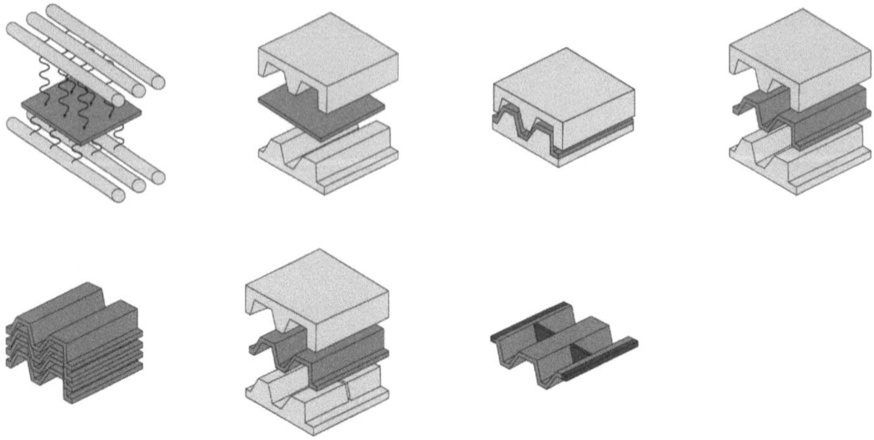

Figure 4.28 Process cycle in a two-step process.

completed once, and a faster process cycle can be achieved [43]. The process cycle runs as follows:

- Creation of a organic sheet cut to size that correlates with the form of the component.
- Preheating and insertion into the special injection molding tool.
- Forming and injection molding of the plastic injection molded component.
- Removal of the component, which does not require post-processing in most cases.

In order to be able to complete preforming in the tool, a movable cavity can be installed for complex structures [47,48]. For example, a rotating tool core that drives out upstream (see Figure 4.29) can make the Ω-shaped contour of the reinforced brace. Needles and support frames must be used to fixate the heated, semifinished product and can be preformed by the closing process. Once the tool closes, injection begins. Figure 4.29 shows the principle of a preforming injection molding tool.

4.5.2.2 Process Sequences

In order to be able to process organic sheets into injection molded parts, several processing steps must be carried out. In general, the machine must have a combination of the following units: injection

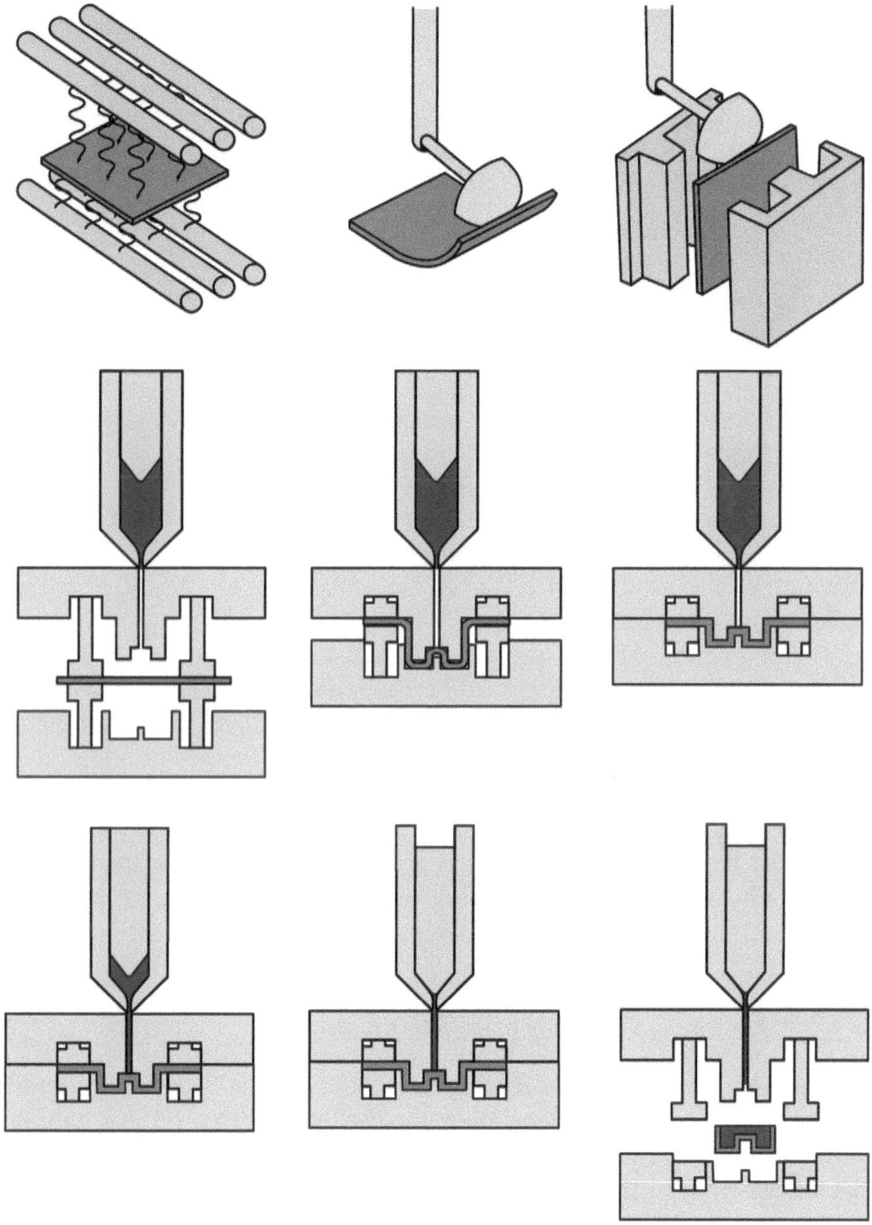

Figure 4.29 Combined thermoforming and injection molding process for the manufacture of components reinforced with organic sheets.

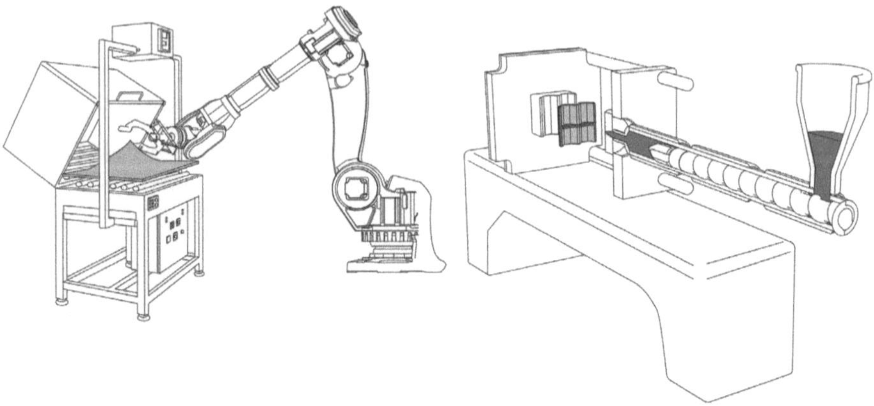

Figure 4.30 Structure of a system for manufacturing organic sheet-reinforced, injection molded parts.

molding machine, preheating station, preforming and injection molding tool, and a handling system. Figure 4.30 provides an example of the structure of the production plant for the processing of organic sheets in injection molding processes.

The relocation of the components is usually carried out by robot systems, which form the interface between the various units. The transfer points during handling of the semifinished products and for communication between the units represent a special challenge. In addition to the smoothly running process cycle, which is suitable for large series owing to its cycle time, the production-induced safety requirements must also be fulfilled. Factors that significantly influence the component quality include the temperature of the organic sheets, the injection speed, and the melt and tool temperatures [45].

4.5.2.2.1 Preparation of Organic Sheets

Varying mechanical processing methods can be used to generate organic sheet cuttings. Depending on the complexity of the cutting, saws and milling machines should be used. This processing step can be decoupled from further steps.

The cutting of the organic sheets must be manufactured very precisely. The semifinished product, which will be inserted into the injection molding tool, cannot have any protruding sections that could get clamped in the parting plane. Cuttings that are too small can get pulled into the contours of the cavity during additional forming in the injection molding tool, and, consequently, may not be able to provide the required reinforcement in the marginal areas. This should be taken into account in advance by actions such as using draping simulations. Fluctuations in the size of the organic sheet, and, thus, in the free volume of the injection molding cavity, lead to divergences in the injection molding volume, which are hard to intercept.

4.5.2.2.2 Preheating

Infrared radiators are generally used to preheat the organic sheets, but other heating systems (i.e., convection heating) are also an option. The temperature of the matrix should be increased until it is possible to carry out a forming process. The matrix material must have softened, and its temperature must be above the glass transition temperature (amorphous thermoplast), or in the range of the crystallite melt temperature (semicrystalline thermoplast). Special attention should be paid to the cooling behavior during repositioning processes, and its influence should be minimized by a quick handling system. The temperature can be monitored without any contact; that is, by means of a pyrometer [49]. Here, low temperatures result in damage in the fibers, incomplete forming, and a poor connection during subsequent injection molding.

The upstream preheating also significantly influences the injection process. Only once a sufficiently high matrix temperature has been reached can a firmly bonded connection between the thermoplastic components be achieved. However, the temperature can neither fall short of nor exceed the processing window [50]. If the temperature is not sufficiently high once the organic sheet has been positioned in the injection molding tool, additional heating is unavoidable. Figure 4.31 shows experimental results of the decrease in temperature after heating an organic sheet, and underlines the importance of employing a fast handling system.

The radiator power and the wavelength should be adjusted to the material. Overheating of the surface should be avoided, just as insufficient heating of the cross section should. The temperature that can be achieved in the organic sheet depends not only on the material itself (matrix material, fiber content, additives like color, and plate thickness), but also on the employed infrared (IR) radiators (wavelength, power, distance, heating time). A change in the temperature curve throughout the cross section and over time should be taken into account. Therefore, the maximum temperature is to be expected on the upper side of the organic sheet directly after the input of energy. Due to downstream handling and preforming, the surface of the semifinished product cools

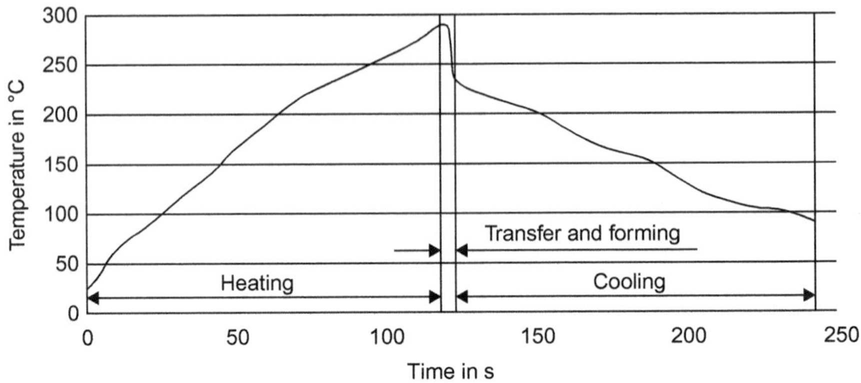

Figure 4.31 Temperature decrease over time after heating a 1.5-mm-thick organic sheet with a PA matrix [51].

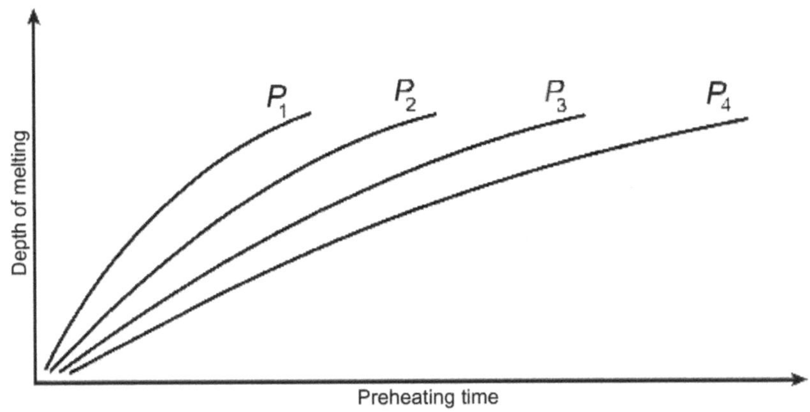

Figure 4.32 Depth of melting in the object to be preheated over time and at different levels of radiator power (power $P_1 > P_2 > P_3 > P_4$).

more quickly, resulting in a temperature maximum inside [50]. In Figure 4.32, a schematic illustration of the correlation between the depth of melting in the object to be preheated and the set radiator power is shown.

Filling materials in the matrix material distribute the radiation, leading to an increased absorption behavior. Thus, an enhanced heating of the surface correlates with the risk of surface damage. Heating inside the component must be achieved by means of thermal conduction. Deeper penetration of the rays is desired, as well as the correlating, enhanced volume absorption. In accordance with the distribution of the intensity $I = I_0 \times e^{-\kappa x}$, the heat balance depends on the absorption coefficient κ of the fiber composite material. Organic sheets with a typically high fiber content and a correlating large absorption coefficient κ hardly enable this. This leads to enhanced surface absorption [52,53].

4.5.2.2.3 Preforming

Preforming is carried out on a press or in an injection molding machine to create the required geometry. For this reason, the organic sheet must be preheated to a temperature range that enables forming. The entire process is significantly shortened by special injection molding tools, which enable forming inside them. Preheating takes place only once to reduce the inevitable cooling directly before insertion, or in the directly opened mold of the injection molding machine [54]. The fiber orientation in materials reinforced with endless fibers is of special importance when developing the design of a component. A rather precisely assessable fiber orientation results from the forming process of the organic sheet. It is decisive for the utilization of the special properties of reinforced components. They are component properties with especially high mechanical values in the direction of fibers and are tailored to the loading conditions. If mechanical loading in a particularly disadvantageous direction occurs in a formed component, the properties decrease significantly because forming an organic sheet is not comparable to deep drawing. Above all, shearing within the fiber fabric, which is usually not elastic, causes the change in form. This is also referred to as the *trellis effect*. The changing angle between the fiber strands causes locally differing mechanical properties, which can be depicted by a so-called draping simulation. Deformations that are too large can create folds and are generally to be avoided [48]. It is important to take the negative, mechanical effects of narrow edge rounding into consideration when forming organic sheets [55].

Figure 4.33 illustrates the influence of the direction of loading on the strength of an organic sheet with fibers aligned in two directions, oriented vertical to one another. A unidirectional structure (orientation of the fibers in one direction) results in an even larger decrease in the properties if loading is exclusively orthogonal [43].

Draping simulations can be used that depict the forming process and the resulting orientations and transfer the model to the injection molding simulation [44,48]. This is especially significant when complex components with three-dimensional (3D) structures must be produced,

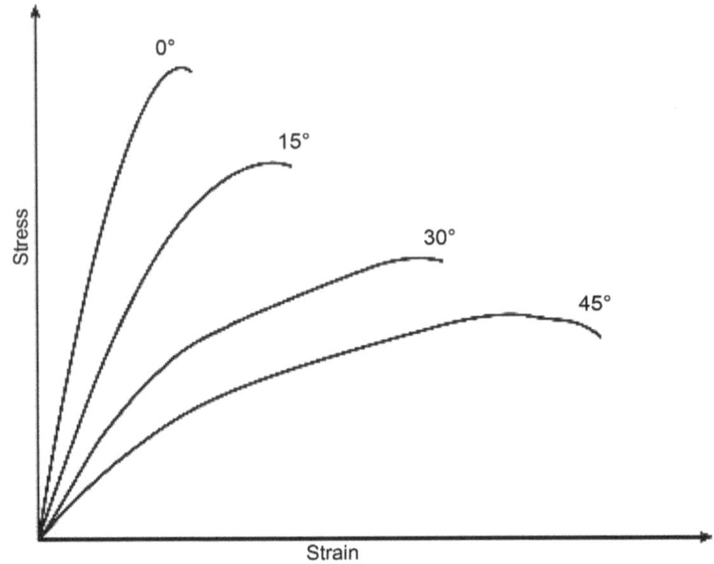

Figure 4.33 Stress–strain diagram of varying fiber loading directions, modeled after [43].

in which a distortion of the original form of the organic sheet takes place; that is, by circumferential reinforcement.

4.5.2.2.4 Injection Molding

Ultimately, the injection molding process does not differ much from methods in which the insert is insert molded or molded. A handling system positions the organic sheet in the open tool. The challenge is securely positioning the already heated, semifinished product into the injection molding tool. Consistent cooling must be considered, and the temperature must not be allowed to fall under critical temperatures, as already mentioned. During the possible preheating process, the tool is still open. After that, it closes.

The cavity is completely filled by the injection of the melt and the holding pressure. Preheating promotes insert molding, and the connection of the endless fibre fabrics. Ideally, the interface between the organic sheet and the injection molded component is hard to identify on the side of the matrix.

Generally, the injection molding tool must be adjusted to process organic sheets. Needle-shaped cores make it possible to position the organic sheets in the cavity without obstructing the closing level, or falling out. Consequently, the contours of the cavity can be neatly molded, the organic sheets are sealed and delamination is counteracted [49].

The essential basics of the injection molding process, and general special features of inserts are not covered in detail here. Injection, the holding pressure phase, and the cooling of the injected form do not differ much from compact injection molding with inserts.

Also, in terms of removal, the manufacturing process for organic sheet-reinforced components does not differ from other injection molding processes. Because the structural components are generally quite large, the finished injection molded part, which does not require postprocessing, is removed by the handling system and set aside.

References

[1] Habenicht G. Kleben. Grundlagen, Technologien, Anwendungen. Berlin, Heidelberg: Springer Berlin Heidelberg; 2009.

[2] Schuck M. Kompatibilitätsprinzipien beim Montagespritzgießen. Universität Erlangen-Nürnberg; 2009.

[3] Gleich H (2004): Zusammenhang zwischen Oberflächenenergie und Adhäsionsvermögen von Polymerwerkstoffen am Beispiel von PP und PBT und deren Beeinflussung durch die Niederdruck-Plasmatechnologie, Universität Duisburg-Essen. Dissertation, Duisburg, 2004.

[4] Pocius AV. Adhesion and adhesives technology—an introduction. München: Hanser Verlag; 2012.

[5] Kuhmann K. Prozeß- und Materialeinflüsse beim Mehrkomponentenspritzgießen. Erlangen-Nürnberg: Lehrstuhl für Kunststofftechnik; 1999.

[6] Kopczynska A. Oberflächenspannungsphänomene bei Kunststoffen—Bestimmung und Anwendung. 1. Aufl. Erlangen: Univ. Erlangen-Nürnberg, Lehrstuhl für Kunststofftechnik (Technisch-wissenschaftlicher Bericht/ Lehrstuhl für Kunststofftechnik, Universität Erlangen-Nürnberg, 52). 2008.

[7] Bischof C, Possart W. Adhäsion—Theoretische und experimentelle Grundlagen. Berlin: Akademie Verlag; 1982.

[8] Fowkes F. Acid–base interactions in polymer adhesion. In: Microscopic aspects of adhesion and lubrication; proceedings of the 34th international meeting of the Société de Chimie Physique, Bd. 7, pp. 119–37.

[9] Bolger JC, Acid base interactions between oxide surfaces and polar organic compounds. In: Adhes Asp Polym Coating (S):3–18.

[10] Lee L-H. Molecular bonding and adhesion at polymer-metal interphases. The Journal of Adhesion 1994;46(1–4):15–38.
[11] Giese VM. Adhäsive Kunststoff-Metall- und Kunststoff-Kunststoff-Verbindungen im Hinblick auf die Spritzgußverarbeitung. Dissertation, Technische Fakultät, Universität Erlangen-Nürnberg, DE; 1995.
[12] Assefpour-Dezfuly M, Vlachos C, Andrews EH. Oxide morphology and adhesive bonding on titanium surfaces. J Mater Sci 1984;19:3626–39.
[13] Poggel H. Untersuchungen zur mechanischen Haftung in Verbundsystemen. Siegen: Höpner und Göttert; 2002.
[14] Byskov-Nielsen J, Balling P. Laser structuring of metal surfaces: micro-mechanical interlocking. Applied Surface Science 2009;255(10):5591–4.
[15] Byskov-Nielsen J, Boll JV, Holm AH, Højsholt R, Balling P. Ultra-high-strength micro-mechanical interlocking by injection molding into laser-structured surfaces. International Journal of Adhesion and Adhesives 2010;30(6):485–8.
[16] Ehrenstein GW, Zhao G. Kunststoff-Metall-Hybridtechnik. ZWF 2001;96 (3):132–7.
[17] Zhao G. Spritzgegossene, tragende Kunststoff-Metall-Hybridstrukturen. Konstruktion, Prozessanalyse und Charakterisierung, Erlangen-Nürnberg; 2001.
[18] Al-Sheyyab A. Light-weight hybrid structures—process integration and optimized performance. Erlangen: Lehrstuhl für Kunststofftechnik; 2008.
[19] Johannaber F, Michaeli W. Handbuch Spritzgießen. 2. Aufl. München Hanser; 2004.
[20] Menges G, Michaeli W, Mohren P. Spritzgießwerkzeuge. Auslegung, Bau, Anwendung. 6. Aufl. München: Carl Hanser Verlag; 2007.
[21] Bichler M. Prozessgrössen beim Spritzgiessen. Analyse und Optimierung, 2. Berlin: Beuth Verlag GmbH; 2012.
[22] Eyerer P, Hirth T, Elsner P. Polymer engineering. Heidelberg: Springer; 2008.
[23] Knappe W. Beeinflussung der Eigenschaften von Kunststoff-Fertigteilen durch die Verarbeitungsbedingungen: VDI, Bildungswerk BW-1734.
[24] Ehrenstein GW. Polymer-Werkstoffe: Struktur—Eigenschaften—Anwendung; Hanser; 1999.
[25] H. Ridder; J. Schnieders: Hybridspritzgieβen-Möglichkeiten und Grenzen, im Tagungsband zur Tagung Spritzgieβen 2007, Oberflächen von spritzgegossenen Teilen, Hybride Bauteile und Elektromechanik, 14.-15.02.2007, VDI Verlag, Düsseldorf; 2007.
[26] Belyi A, Schtscherbakow SW, Pessezki SS. Zur Untersuchung spritzgegossener Adhaesionsverbunde Polyamid—Metall. Plaste und Kautschuk, 1980; p. 642–646.
[27] Haufe, A. (1996): Untersuchung zur Vorhersage der Bindenahtfestigkeit in spritzgegossenen Formteilen: Technische Hochschule Darmstadt. Dissertation; 1996.

[28] García Gutiérrez MC, Rueda DR, Balta Calleja FJ. Microhardness study across the weld line in doubly injection molded glassy polymers. J Mater Sci Lett 1999;18:1237–8.
[29] Ridder H, Heim H-P, Jarka S. Möglichkeiten und Grenzen variabler Werkzeugtemperierung. Kunststoffe 2009;99(5):22–9.
[30] Thinel P, Broer E, Vitz C, Sonderteil I. Oberflächenfehler an thermoplastichen Spritzgiessteilen. Plastverarbeiter, Heidelberg 1995; 46: p. 92–97.
[31] Ehrenstein, G.W.; Ahlers-Hestermann, G. (2004): Handbuch Kunststoff-Verbindungstechnik: Hanser. München Online verfügbar unter <http://books.google.de/books?id = 2sfHdnym7gYC>.
[32] Grujicic M, Arakere G, Pisu P, Ayalew B, Seyr N, Erdmann M, et al. Application of topology, size and shape optomization methods in polymer metal hybrid structural lightweight engineering. Multidiscipline Model Mater Struct 2008;4.
[33] Grujicic M, Sellappan V, Arakere G, Seyr N, Erdmann M. Computational feasibility analysis of direct-adhesion polymer-to-metal hybrid technology for load-bearing body-in-white structural components. J Mater Process Technol 2008;195(1–3):282–98.
[34] Grujicic M, Sellappan V, Omar MA, Seyr N, Obieglo A, Erdmann M, et al. An overview of the polymer-to-metal direct-adhesion hybrid technologies for load-bearing automotive components. J Mater Process Technol 2008;197(1–3):363–73.
[35] Grujicic M, Sellappan V, Pandurangan B, Li G, Vahidi A, Seyr N, et al. Computational analysis of injection molding residual-stress development in direct-adhesion polymer-to-metal hybrid body-in-white components. J Mater Process Technol 2008;203(1–3):19–36.
[36] Geiger M, Ehrenstein GW. Werkstoff- und prozessoptimierte Herstellung flächiger Kunststoff-Kunststoff und Kunststoff-Metall-Verbundbauteile. Tagungsband zum Berichts- und Industriekolloquium 2003 des SFB 396. Maisenbach Bamberg: Universität Erlangen-Nürnberg; 2003.
[37] Malek T. Leichter als Stahl. Hybridtechnik. Kunststoffe 2010;(3):80–4.
[38] Opperbeck G, Raulf R, Zhao G. Gewichtsoptimiert und torsionssteif. Kunststoffe 2008;(3):78–9.
[39] Risthaus M. Eine dauerhafte Verbindung. Kunststoffe 2010;(3):85–8.
[40] Steinbichler G, Rettenwander T. Zukünftige Chancen für den Spritzgießer bei der Verarbeitung thermoplastischer Faserverbundkunststoffe. In: Fachtagung Thermoplastische Faserverbundkunststoffe 2013; (2), S: 135–45.
[41] Bonefeld D, Giehl S, Haspel J, Jäschke A, Lahr R, Obermann C, et al. Kombination von Thermoplast-Spritzguss und Thermoformen kontinuierlich faserverstärkter Thermoplaste für Crashelemente (SpriForm). Gemeinsamer Schlussbericht zum BMBF-Verbundprojekt. Förderkennzeichen: 02PU2350-02PU2355. BMBF-Forschungsbericht. Bundesministerium für Bildung und Forschung; 2012S: 1–52.

[42] Kunststoff Magazin. Alternative zu Aluminium und Stahl. KunststoffMagazin 2006;(6):64–9.
[43] Lutter F, et al. Leicht und hoch belastbar. In: Kunststoffe, Themendossier Leichtbau 2011;(11), S: 14–17.
[44] Brandt M, et al. Integrative simulation von organoblech-hybridbauteilen. Kunststoffe, Hanser Verlag 2012;(9), S:102–6.
[45] Moritzer E, Budde C, Tröster T, Pöhler S. Parametereinfluss auf die Verbundfestigkeit einer Organoblech-Kurzfaserthermoplast-Werkstoffkombination. Joining Plastics—Fügen von Kunststoffen 2014;(2):100–5.
[46] Lahr R, Weick T, Mitschang P. Neues Fertigungskonzept für die Formnestumformung bei thermoplastischen CFK-Bauteilen. In: IVW-Schriftenreihe, Institut für Verbundwerkstoffe, Kaiserslautern 2004;(Band 48), S: 317–22.
[47] Schuck M. Großserientauglich, leicht und belastbar. Hybridverfahren Spriform mit Organoblech für den Automobilbau auf dem Vormarsch. Industrieanzeiger 2010;Band 132(18):28–9.
[48] Zimnol R, Lutter F, Malek T, Arping T. Chancen der Organoblech-Hybridtechnik im Leichtbau. Automobiltechnische Zeitschrift - ATZ 2012; Band 114(3):210–15.
[49] Bonefeld D, et al. Gemeinsamer Schlussbericht zum BMBF—Verbundprojekt: Kombination von Thermoplst-Spritzguss und Thermoformen kontinuierlich faserverstärkter Thermoplaste für Crashelemente (SpriForm). Förderkennzeichen: 02PU2350 - 02PU2355. 2011.
[50] Egger P, Stock A. Vollautomatisierte Fertigung von Kunststoff-Faserverbundbauteilen. Spritzgießen, Schriftenreihe Kunststofftechnik (VDI) 2011;(Band 4314), S: 151–64.
[51] Junge, M. (2013): Gleitziehbiegen von faserverstärkten Thermoplasten. Lehrstuhl für Umformtechnik, Universität Siegen. Available from: http://www.mb.uni-siegen.de/uts/.
[52] Throne JL, Beine J. Thermoformen: Werkstoffe, Verfahren, Anwendung. München, Wien: Hanser; 1999.
[53] Pedrotti F, et al. Optik für Ingenieure—Grundlagen. Berlin, Heidelberg: Springer Verlag; 2005.
[54] Malek T, Obermann C. Gewusst wie—richtig konstruieren. Hybridbauteile aus thermoplastischen Hochleistungsverbundwerkstoffen. Der Konstrukteur 2013;(6): 12ff.
[55] Schmachtenberg E, Al-Sheyyab A. Kurzer Prozess bei hybriden Strukturen. In: Kunststoffe, Hanser Verlag 1997; 97 (12), S: 120–4.
[56] ENGEL AUSTRIA GmbH, Schwertberg, Austria, published at < www.engelglobal.com >, found on 01.06.2015.

5 Injection Molding of Bio-Based Plastics, Polymers, and Composites

Maik Feldmann and Johannes Fuchs
Universität Kassel, Institut für Werkstofftechnik, Kunststofftechnik, Kassel, Germany

5.1 Introduction

Plastics made from renewable raw materials represent an alternative to conventional plastics based on fossil resources. In the past, strong growth of the production capacity of bio-based, nonbiodegradable plastics has been observed [1]. From 2008 to 2012 alone, the production capacities multiplied by a factor of 4 (Figure 5.1).

Plastics manufactured from renewable raw materials are often referred to as bioplastics or biopolymers. These terms are repeatedly a source of confusion because plastics exist that are biodegradable, and

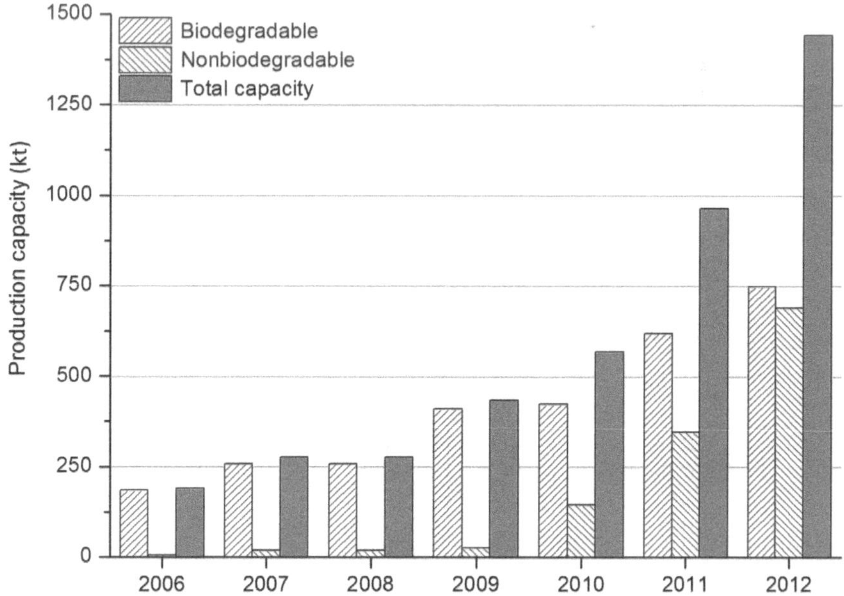

Figure 5.1 Worldwide production capacities of bioplastics according to Ref. [1].

Specialized Injection Molding Techniques.
DOI: http://dx.doi.org/10.1016/B978-0-323-34100-4.00005-5

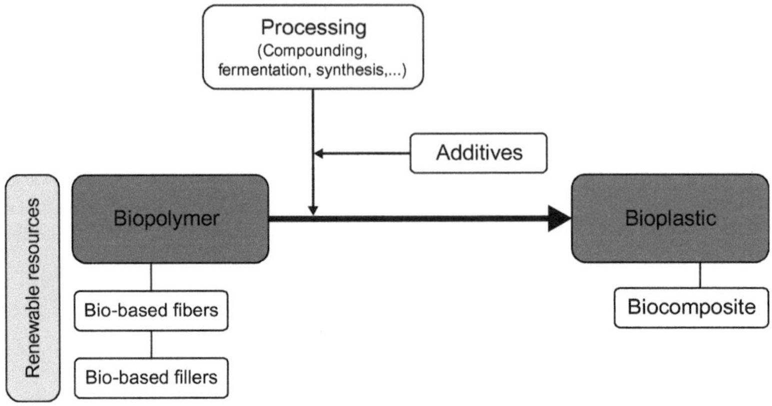

Figure 5.2 Schema of the linkage between bioplastic and biopolymer.

therefore are associated with the word bioplastics. However, they are not necessarily bio-based. Bio-based plastics consist of renewable raw materials and can be biodegradable, but they do not have to be. Bio-based plastics can be synthesized, either entirely or partially, from renewable raw materials. Moreover, natural biopolymers can be directly used in plastics processing.

Biopolymers are the basis of bio-based fibers and fillers, and they are also the raw material basis for bioplastics. Only plastics that are partially or entirely bio-based are called bioplastics. Only bio-based biodegradable plastics are referred to as bioplastics; petro-based biodegradable plastics are not. Biocomposites are made of a bioplastic and a bio-based fiber, a bio-based filler, or both.

In order to communicate a better understanding of the previously mentioned terms, the correlations have been depicted in Figurc 5.2.

Until now, bioplastics have primarily been used for short- to medium-term applications. These include packaging and disposable products like plastic bags or cups (see Figure 5.3). Conventional processing methods, for example, include thermoforming and film blowing. In injection molding, very few applications are known so far.

5.2 Materials

Different kinds of bio-based polymers and bio-based plastics are available nowadays. The range of properties is just as extensive as for common plastics and composites. Before diving into an in-depth elaboration, the terminology will be defined precisely next.

Figure 5.3 Examples of the various uses of bio-based composites in injection molding. Top left: Automobile radiator part made of bio-based PA with 20 wt% cellulosic fiber; injection molding machine: Krauss Maffei 200–900C1, shot weight 240 g. Top right: Cover made of foamed PLA with 50 wt% TPS; injection molding machine: Engel Duo700 MuCell®, shot weight 180 g. Bottom: Flower bowl made of PLA with 50 wt% potato starch; injection molding machine: Engel Duo1500, shot weight 3500 g.

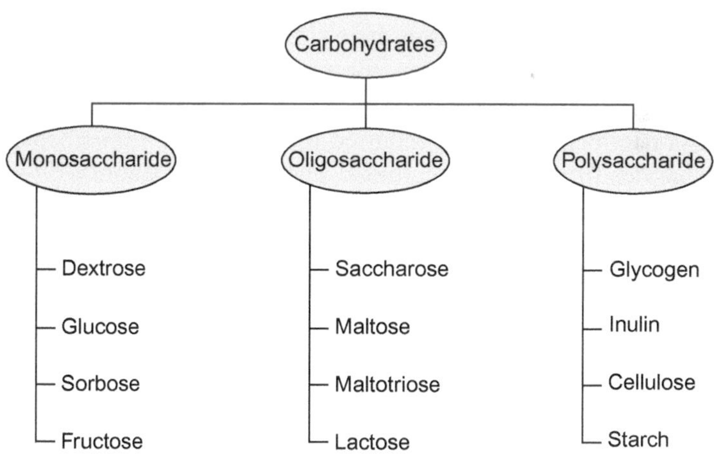

Figure 5.4 Classification of carbohydrates according to Ref. [2].

The polysaccharides starch and cellulose make up the largest proportion of biopolymers. Besides being used in fermentation processes, starch can be directly used. Starch ($C_6H_{10}O_5$) belongs to the polysaccharides, which in turn belong to the material class of carbohydrates (Figure 5.4). During photosynthesis, plants produce starch and store it as an energy reserve. Depending on the type of plant, starch is stored in different areas. Potatoes, arrowroot, and manioc store starch in their spuds/bulbs and roots. In contrast, wheat, corn, and rice store starch in their seeds. Other plant types deposit the starch in their fruits (i.e., peas), or in the marrow (i.e., sago palm) [3].

Biopolymers, or native polymers, are synthesized by means of processes within the plant when using renewable resources. In addition to the synthesis energy provided by the Sun, carbon dioxide from the atmosphere is used for the manufacture of the biopolymer. Both cellulose and starch can be incorporated into plastics as reinforcement or filling materials once they have been extracted from the plant [4]. In order to process biopolymers in thermoplast processing (i.e., starch) using conventional machines, a modification is needed, such as by adding a plasticizer (water, glycerin, etc.).

Bioplastics made from renewable raw materials require an additional synthesis step or chemical modification that does not take place naturally. Examples include the following bioplastics: (i) polylactic acid (PLA), (ii) polyhydroxy butyric acid (PHB), (iii) thermoplastic starch (TPS), and (iv) polyamides type X.10. Various basic raw materials are suitable for the synthesis of bioplastics, like cornstarch, cane sugar, and castor oil. The chemical structure of the polymer determines whether the plastic will be biodegradable.

The so-called drop-in solutions are a subgroup of bioplastics. In contrast to the manufacture of entirely new plastics like PLA, already-known plastics can be made from renewable raw materials instead of fossil resources by means of fermentation. The required monomers are implemented using polysaccharides from bacteria cultures. These drop-in solutions, regardless of the raw material sources, generally display the same properties as their conventional counterparts. Bio-based PA 6 and PA 6.6 belong to them and are created from fermented adipic acid and caprolactam. Moreover, various types of bio-based polyethylene are commercially available.

5.2.1 Bioplastics

5.2.1.1 Thermoplastic Starch (TPS)

In order to manufacture TPS, native or modified starch is destructured thermomechanically in an extrusion process. Employing a related screw configuration with an adequate number of kneading elements, as well as a sufficiently high temperature, is very significant. When manufacturing TPS, corotating twin screw extruders are used. During compounding, in addition to water, different plasticizers (i.e., glycerin, sorbitol, and urea/carbamide) are added to induce chemical destructuring in addition to the thermomechanical destructuring of the starch. This enables the specific adjustment of individual properties of the starch, such as its stiffness [5–7].

The degree of destructuring can be quantified using microscopic, radiographical, thermoanalytical, and differential scanning calorimetry (DSC) analyses. In the case of thermoanalysis using a DSC measuring instrument, an increase of the gelatinization enthalpy indicates an increased destructuring of starch. In general, each native starch can be processed into TPS. However, depending on the botanical origin of the starch, differences in the properties can be seen. For example, there can be a difference in the amount of energy needed in order to destructure the native starch entirely. In the case of potato starch, approximately 650 kJ/kg are needed; in contrast, the destructuring of wheat starch requires only about 435 kJ/kg, and cornstarch only 380 kJ/kg [8]. TPS generally displays the properties of an amorphous thermoplast and can be processed using conventional plastic-processing machines. As in the case of native starch, TPS possesses a hydrophilic character that limits its areas of application. The mechanical and optical properties of a component or sample made of TPS are influenced by the moisture content. Specific conditioning of TPS before injection molding can be used to induce foaming in the material. Due to the hydrophilic properties of native starch and TPS, these biopolymers are often blended with other plastics to enhance their properties and broaden their range of application [9–13].

5.2.1.2 Starch Blends

Starch blends represent the largest proportion of bioplastics on the market by far. In literature, starch blends are classified as a mixture of TPS with other bio-based plastics. In contrast, starch-filled plastics are often considered as a mixture of native starch and another bio-based plastic. Depending on whether the matrix polymer or the TPS are inserted via the main feed or the side feeder of the extruder, the two-phase mixture represents either the continuous phase or the dispersedly distributed secondary phase of the starch blend. In the compounding process, the physical mixing creates a bond between the TPS and the associated matrix polymer.

By utilizing many kneading elements, both the mixture and the incorporation of the starch in the matrix polymer can be enhanced. However, it is important to consider the fact that using shearing forces that are too strong leads to an additional destructuring of the starch. Thereby, the particle size and particle distribution of the starch grains shrink significantly [12,14–18]. The tensile properties of selected compositions (Table 5.1) are shown in Table 5.2, providing an example of the properties of various starch blends and matrix materials [19] (Figure 5.5).

Table 5.1 Investigated Materials

Product name	Note	Manufacturer
PP 575P	Polypropylene	Sabic
PLA 3051D	PLA	NatureWorks
PLA 2002D	PLA	NatureWorks
Licocene® PP MA 6452	Coupling agent	Clariant
Superior®	Native potato starch	Emsland Group
Bioplast® TPS	TPS (modified composition)	Biotec

Table 5.2 Tensile Strength and Young's Modulus of the Tensile Tests for Selected Starch Blends and Their Matrix Plastics in Accordance with ISO 527

Material	Tensile Strength, σ_m (MPa)	Young's Modulus (MPa)
TPS PLA	50.4	2635
PLA starch	52.2	4204
PLA 3051D	54.8	2186
PP starch	12.1	1975
PP starch MA-PP	16.5	2027
PP 575P	25.4	1205

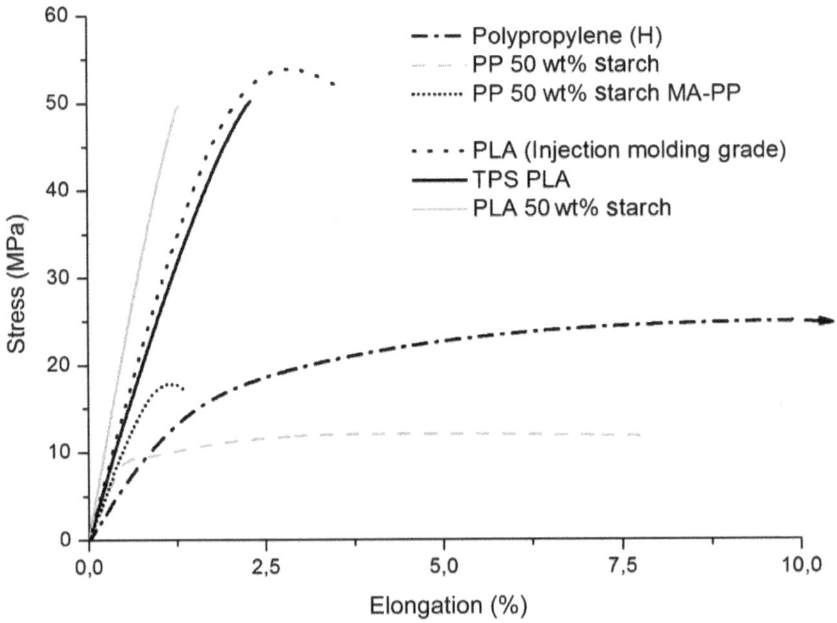

Figure 5.5 Stress–strain diagram of selected starch blends and their matrix plastics.

5.2.1.3 Biopolyamides with a Vegetable Oil Basis

Vegetable oils, such as castor oil, are another raw material that can be utilized for the synthesis of bioplastics. The fatty acids contained in the vegetable oils can be used as the basic module for polymers. Here, technical bioplastics, such as polyamide (PA) 6.10, present themselves as an alternative to the available petrochemical polymers, and therefore, they have become a central focus. When combined with natural fibers, these bioplastics can be used in technical areas with higher mechanical stresses, but thermal ones as well (see Section 5.2.3). In addition to PA, there are other technical plastics based on renewable raw materials, such as polybutylene terephthalate (PBT), polyurethane (PUR), polyethylene terephthalate (PET), and polytrimethylene terephthalate (PTT). Yet, only some of them are partially bio-based [20].

Owing to their significance in technical applications and their comparably good availability in contrast to other technical bioplastics, biopolyamides will be discussed here in more detail. Unlike PAs based on crude oil, biopolyamides differ in terms of the origin of their raw materials, and, as is common for PAs, with regard to their chemical makeup. The most well known raw material basis is that of the castor plant. The fruits of the castor plant contain approximately 40–50% castor oil, which, in turn, consists of approximately 90% ricinoleic acid ($C_{18}H_{34}O_3$). The ricinoleic acid is split into a C_8 and a C_{10} molecule by means of alkaline hydrolysis. The more valuable C_{10} molecule can be used to synthesize sebacic acid as a monomer for the type X.10 PA. The C_8 molecule is deployed, for example, in plasticizers [21–23].

The best-known and most commercially available biopolyamides are PA 6.10, PA 10.10, PA 11, and PA 4.10. Some of the biopolyamides are entirely based on renewable resources, and others are partly bio-based (see also Table 5.3). Furthermore, there are also completely or partially bio-based types that are either not available at all or only selectively commercially available. At present, nearly every large PA manufacturer offers one or several bio-based PAs in their product portfolio under various trade names. The bio ratio is listed in wt% and refers to the carbon atoms contained in the polymer. In this observation, the origin of other atoms is not taken into account [20].

Nearly all bio-based PAs are available with the associated reinforcement fibers (i.e., glass fibers). Bio-based PAs are already used in cable and tube manufacturing.

Table 5.3 Selected Properties of Various PAs [20]

PA Type	Ratio of Renewable Raw Materials (%)	$-CH_2/NHCO$ Ratio	Melting Point (°C)	Density (g/cm^3)	Water Absorption (%)	
					23°C, 50% Relative Humidity	Saturated
PA 4.6	0	4	295	1.21	3.8	12
PA 6.6	0	5	260	1.15	2.8	8–9
PA 6	0	5	222	1.14	3	9–10
PA 6.10	~63	7	220	1.07	1.4	3.3
PA 6.12	0	8	215	1.06	1.3	2.8
PA 10.10	~100	9	200	1.44	1	1.8
PA 11	~100	10	185	1.03	1.1	1.8
PA 12	0	11	178	1.01	0.7	1.5

5.2.1.4 Polylactic Acid

The basic constituent for the synthesis of PLA is native starch, meaning a biopolymer. In the first step, native starch is fermented with the aid of microorganisms by means of enzymatic hydrolysis. Subsequently, the created lactic acid is synthesized into lactide and later polymerized into PLA [24]. A bioplastic is created that is biodegradable thanks to its chemical structure and is mostly transparent. In comparison to conventional thermoplasts, PLA displays several characteristics with regard to its property and processing spectra, including low heat distortion temperature, low hydrolysis resistance, and slow crystallization speed. The synthesis procedure of the bioplastic PLA is shown in Figure 5.6.

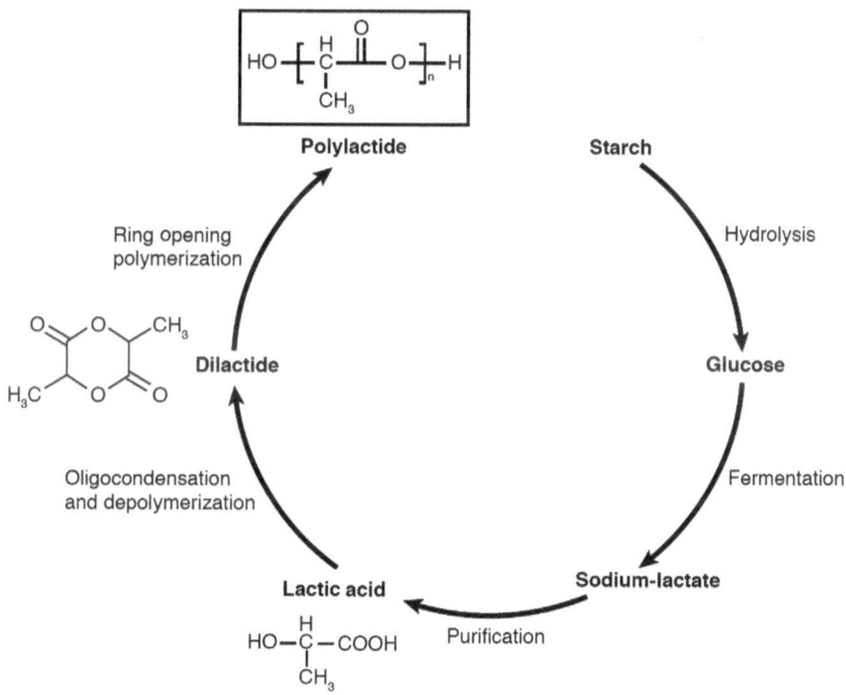

Figure 5.6 Synthesis of PLA from starch according to Ref. [25].

5.2.2 Cellulosic Fibers

Natural fiber-reinforced plastics represent a separate group among the fiber-reinforced plastics, and are called biocomposite materials. Here, duroplastic and thermoplastic matrices can be selected. However, in this chapter, the focus will be on natural fiber-reinforced plastics with a thermoplastic matrix. Moreover, natural fibers represent an inexpensive

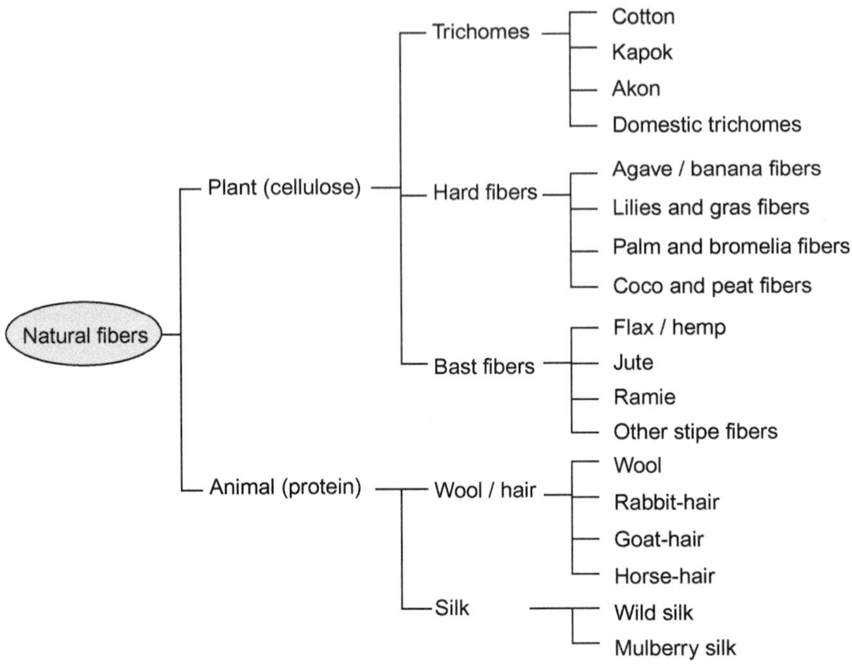

Figure 5.7 Classification of natural fibers in accordance with DIN 60001 [27].

alternative to glass fibers. Owing to the previously mentioned advantages of natural fiber-reinforced plastics, the demand on the market for technical applications made of these materials is continuously increasing [4,24–26].

Since the beginning of the 1990s, natural fiber-reinforced plastics have contributed to reduce the amount of CO_2 and to sustainability. On the one hand, this is due to their lightweight potential, which results from having a lower density than that of glass fiber-reinforced plastics. On the other hand, it is also due to their individually adjustable properties in technical applications [24]. The biogenic proportion equals the fiber content in the composite. When using a bio-based matrix, natural fibers enable the realization of completely bio-based composites.

In general, the term natural fibers refer to those that are manufactured from a natural raw material like plants, animals, or minerals. For example, these natural or native fibers can be extracted from hemp, jute, flax, or wood. Plants that are suitable for fiber extraction can be grown in nearly every climatic area of the Earth. This enables transport distances to be kept short and costs kept low. However, fluctuations do arise with regard to the quality of plant-based raw materials, and these can affect the mechanical properties of the substances in question. The various natural fibers are differentiated according to their raw material source and fiber type (see Figure 5.7) [27,28].

Table 5.4 Mechanical Properties of Selected Cellulosic Reinforcement Fibers [29–33]

Fiber	Density (g/cm^3)	Tensile Strength (MPa)	E-Modulus (GPa)	Elongation at Break (%)
Jute	1.3–1.45	393–773	13–26.5	1.16–1.5
Flax	1.5	345–1100	27.6	2.7–3.2
Hemp	1.5	690	N/A	1.6
Ramie	1.5	400–938	61.4–128	1.2–3.8
Sisal	1.45	468–640	9.4–22	3–7
Abaca	1.5	500–764	6.2–31	1–7
Man-made cellulose Cordenka 700S3	1.5	833	20	13

The group of plant-based natural fibers is of special interest as a reinforcement fiber. Animal fibers can also be used as reinforcement fibers in plastics. The most well established mineral natural fiber that was used in technical contexts in the past was asbestos. Because of its health risk, it is no longer used significantly. The mechanical and physical properties of the various but similar natural fibers essentially depend upon their chemical and structural properties (see Table 5.4).

Plant-based natural fibers themselves are a composite material that consists of various components. Cellulose, lignin, and hemicellulose are the main constituents. These components are contained in the natural fibers in various proportions, and each has a specific function in the plant fiber.

In the case of bio-based reinforcement fibers, man-made cellulose fibers (viscose) are particularly special. It is synthetically manufactured from the renewable raw material cellulose. For many years now, these fibers have been used for reinforcing e.g., high performance tires. The chemical process leads to unvarying fiber qualities, which are comparable to those of glass fibers. Different processes for the manufacture of chemical fibers based on renewable raw materials are known, which decisively influence the mechanical properties of the fibers. The viscose method, which employs carbon disulfide, is of utmost industrial importance. Unlike native natural fibers, the mechanical properties of regenerate fibers can be strongly influenced by the manufacturing process, the crystallization degree, and the orientation. A growth-induced fluctuation of the properties and irregular fiber diameters, which can occur in natural fibers, are not present in man-made cellulose fibers [20,21].

An essential disadvantage of cellulosic reinforcement fibers is their low thermal resistance in comparison to conventional reinforcement

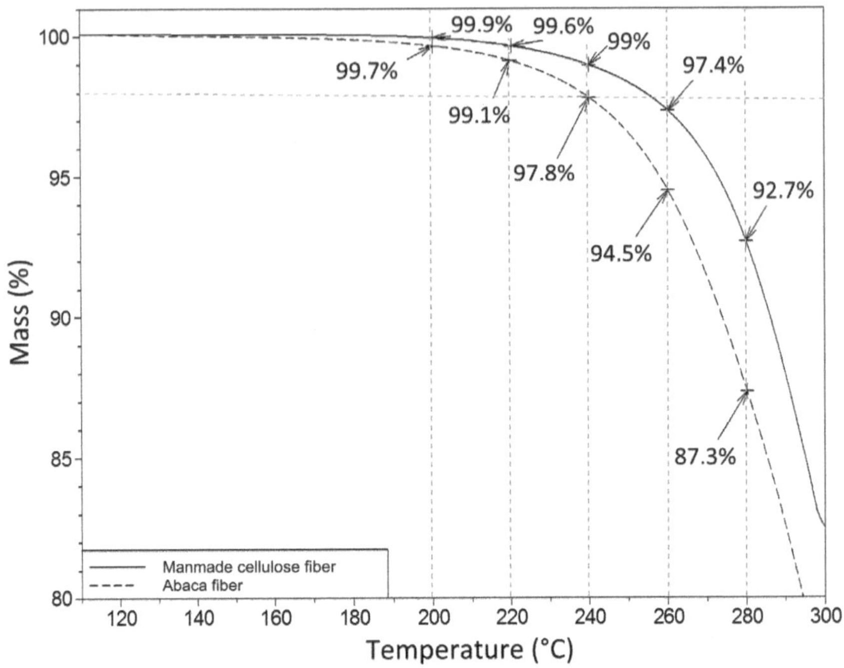

Figure 5.8 Mass loss of selected cellulosic reinforcement fibers at different temperatures using a TGA [20].

fibers like glass fibers. This is of particular significance when using thermoplastic matrix materials with higher melting points. It is known from the research literature that of the three main components of plant-based fibers, lignin and hemicellulose display the lowest heat resistance. This can be explained by the higher molecular structure (crystalline organization) of cellulose compared to hemicellulose and lignin. The degradation of the fibers becomes apparent due to the development of odor, which increases analogous to the temperature. It is also discernible with regard to the tear strength of the fibers, which decreases as the temperature rises [20,28,31,34–36]. Figure 5.8 shows the mass reduction during the thermogravimetric analysis (TGA) measurement of man-made cellulose fibers and abaca fibers depending on the temperature. Here, the significantly lower degradation of man-made cellulose fibers due to the high proportion of cellulose is remarkable.

5.2.3 Starch as a Filling Material

In addition to natural fibers that are used as reinforcement components, filling materials made of renewable raw materials are used in plastics processing. In contrast to the fibers, the filling materials do not

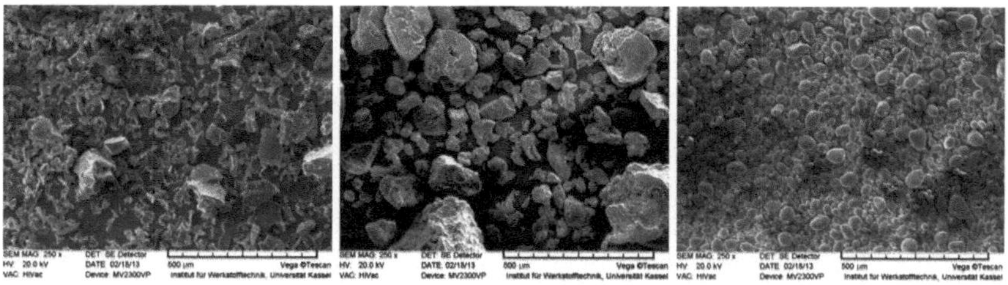

Figure 5.9 SEM images of various starch types [Emjel (left), Empre (center), and native starch (right)].

Table 5.5 Composition and Properties of Selected Starch Types

Type of Starch	Starch Content (%)	Amylose Ratio (%)	Degree of Crystallinity (%)	Grain Size (μm)
Rice	89	14–32	38	3–8
Corn	71	20–30	N/A	5–25
Wheat	74	20–28	36	20–40
Potato	82	20–27	25	20–185

have a definite length–diameter ratio (i.e., aspect ratio). Filling materials are deployed not only to specifically modify the property spectrum, but also to increase the biogenic ratio and reduce costs. Wood and starch particles are used the most.

The starch particles or starch grains (see Figure 5.9) mainly consist of two glucans called amylose and amylopectin (see Table 5.5). There is also a 5–10% proportion of amylopectin C, which displays a mixture of properties of amylopectin and amylose. Depending on the type of starch, the starch grains also contain up to 20% water, various minerals, proteins, lipids, and phosphoric acid. Essentially, starch grains consist of approximately 20–30% amylose and 70–80% amylopectin. However, there are also starches available on the market with a amylopectin ratio of nearly 100%. These are so-called waxy starches (e.g., waxy corn/waxy maize) [37–40].

In addition to native starches, there are destructured starches, including the pregelatinized starch Emjel and the extruded starch Empre (see Figure 5.6).

5.2.4 Wood as a Filling Material

Wood particles that are incorporated into thermoplastic plastics are made of either soft or hard wood and have different particle sizes depending on the preparation process. Initially, production waste materials from wood processing companies were used. Nowadays, wood particles are sometimes specifically produced. One of the best-known products in which wood particles are used are wood polymer composites. They often are used in decking and simple injection molded components.

When processing, it is essential to keep in mind that wood and starch particles are hydrophilic and thermally sensitive like natural fibers. In combination with a thermoplastic matrix, entirely new materials with specific characteristics are created.

5.2.5 Biocomposites

The term biocomposite has not yet been clearly defined. In this book, this term means a matrix made of fossil- or bio-based raw materials that is reinforced or filled with fibers or particles made of renewable raw material. The compounding process must fulfill special requirements for biocomposites. For this reason, various conventional, but also optimized or newly developed, methods are used. In particular, the dosing of cellulosic fibers proves to be a challenge for processing technology when manufacturing natural fiber-reinforced thermoplasts. Continuous and discontinuous methods are used. However, the method decisively influences the resulting mechanical properties of the composite material. In particular, the mechanical and thermal energy application in the material must be considered for suitability in the field of preparing natural fiber-reinforced plastics. If it is too high, chemical degradation of the fibers or a shortening of the fiber lengths occurs; both have negative effects on the composite properties. Research has shown that pultrusion methods are especially suitable for matrix materials with higher melting points when it is desired to mix in cellulosic reinforcement fibers carefully [20,41,42].

Primarily, extrusion methods are used to manufacture biocomposites made of thermoplastic plastics and starch or wood particles. In most cases, the biocomposites are compounded or prepared with filling materials using a corotating twin screw extruder. The corotating twin screw extruder is used frequently due to its excellent mixing effects. The material is also exposed to high shearing forces, which leads to a grinding of the filling materials, sometimes to the point of damage. In this method, the particles are continuously fed into the extruder by gravimetric dosing. The configuration of the dosing sequence is often set so that the thermal degradation of the bio-based filling material is minimized. Thermal degradation is connected to a reduction of the optic and mechanical properties. In some cases, special extruders are utilized, like the planetary gear extruder, other multiscrew extruders, and the conical twin screw extruder. Such machines can be advantageous when preparing sensitive materials.

As a rule, gentle processing should be selected when compounding biocomposites. Such processing includes a short residence time, a gentle screw configuration, low processing temperatures, and, if needed, a predrying of the materials. Otherwise, damage can occur that would lead to negative effects during the subsequent injection molding process due to renewed exposure of the material to thermal stress.

5.3 Specific Injection Molding Behavior

5.3.1 Characteristics in Injection Molding

5.3.1.1 Bioplastics

Bio-based polymers, such as PLA or PHB, are very sensitive to hydrolysis and the degradation of molecular weight caused by thermal stress. One reason is the small range between the melting temperature and the degradation temperature of some bioplastics, like PHB. Also, demolding can be difficult depending on the polymer because of a very low crystallization rate. In particular, PLA displays a very slow crystallization rate compared to other common polymers. Numerous studies focus on the modification of PLA for the improvement of the processing and material properties. For example, Harris and Lee [43] verified significant increases in the degree of crystallinity and the crystallization speed as a result of the nucleation effect of added talcum or ethylene bis stearamide. Due to this effect and the higher level of crystallinity, the thermomechanical properties can be further improved. Tempering the tool also has a significant influence, and the temperature should be set very high when processing PLA in order to achieve a high degree of crystallinity here.

However, this contradicts the generally targeted short cycle time. As a consequence, typically in parts the PLA is in an amorphous state. It is important to take into account that the injection molding tool is designed to allow good demolding. Tools with a polystyrol construction are especially suitable.

On the whole, modifications and processing adjustments must be made when processing bioplastics, so as to be able to compete with conventional plastics both economically and in terms of the material properties. Processing adjustments primarily encompass temperature control, residence time, and shearing.

5.3.1.2 Biocomposites

When processing natural fiber-reinforced plastics, the previously mentioned processing adjustments must be considered. In particular, the moisture absorption of the fibers and the thermal sensitivity (see Figure 5.5 [20]) should be noted since they lead to lower mechanical properties, discolorations, odor development, and processing fluctuations. Extensive exposure to thermal stress (e.g., while in the plastification unit of an injection molding machine) should be avoided because it can result in charring on the surface, which in turn affects the mechanical properties negatively. In a natural fiber-reinforced thermoplast, the matrix material should have a relatively low melting point (lower than the degradation temperature of the deployed filling material/reinforcement material), in order to expose the natural fiber to the least amount of thermal stress possible. The processing parameters should be selected accordingly, so that the cycle and residence times of the material are as short as possible.

In contrast to glass fibers, natural fibers absorb approximately 10% water due to their hydrophilic properties. This, in turn, makes drying natural fiber-reinforced plastics necessary prior to processing using injection molding.

Glass fiber-reinforced plastics have an abrasive effect on the tool surface or machine parts, which can lead to increased wear and ultimately affect the quality of the manufactured products. This effect is decidedly weaker when processing natural fibers.

Essentially, conventional injection molding machines can be used to process natural fiber-reinforced plastics. Moreover, several manufacturers offer specifically configured plastification screws and machine elements. When constructing injection molding machines, it is important to make sure that there are no blind spots or extreme cross-sectional changes that could lead to local irregularities. The layout of hot-runners can be transferred from thermally sensitive polymers and must be adjusted to the comparably high viscosity of natural fiber-reinforced composites.

5.3.2 Process-Induced Material Properties

As previously described, processing bioplastics and biocomposite materials using injection molding makes it necessary to adjust the processing parameters. In particular, the temperature, hydrolysis, and shear sensitivity of some bioplastics must be taken into account. However, contradictory results with regard to these special properties can be found in various studies. Researchers were unable to verify a significant

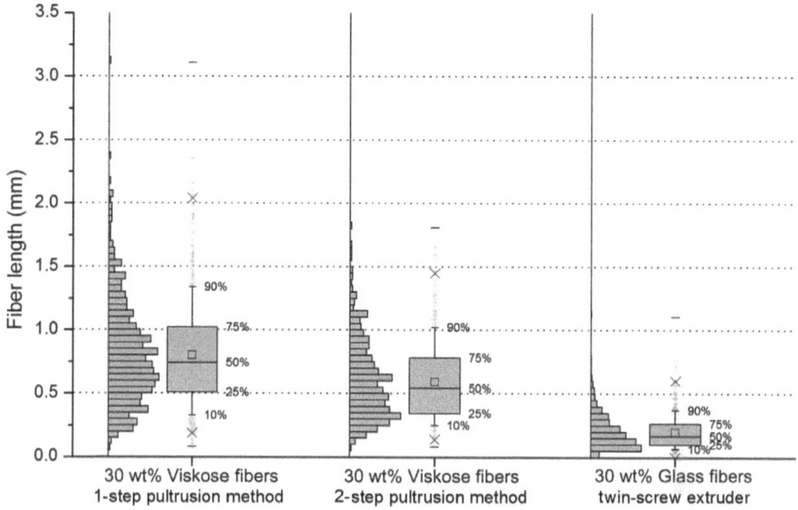

Figure 5.10 Fiber length of PA 6.10-composites injection molded specimen with various fibers and processed using different compounding methods according to Ref. [41].

shearing and hydrolysis sensitivity of PLA in comparison to conventional polypropylene (PP) and polyethylene (PE) in a standard injection molding cycle for the manufacture of type 1A samples (ISO 527-2).

Yet, if the residence time in the cylinder is extended, a higher degree of degradation of PLA occurs with a correlating reduction of the viscosity. Similarly, this is also the case for processing thermoplasts with native filling materials, like starch or wood flour. Consequently, depending on the application and machine, a material-specific, optimal processing window must be determined.

It is important to take the special features of the characteristics of drop-in bioplastics during processing into account, as was the case for bioplastics based on fossil raw materials. Thus, it is crucial to keep the hydrolysis sensitivity and moisture absorption of PA in mind.

When processing thermoplasts reinforced with cellulose, the influence of various preparation methods becomes evident in the distribution of fiber lengths in the injection molded samples. This can be seen in the example of cellulose fiber-reinforced PAs (Figure 5.10). The thermomechanical damage leads to an uneven distribution of fiber lengths, and thus also influences the mechanical properties. Also, when compared to glass fiber-reinforced plastic, a significant difference is discernible. In comparison to cellulosic reinforcement fibers, much shorter fiber lengths are in the component. This can be explained by the higher elongation at break and the lower stiffness of the cellulose fibers (compare Table 5.4).

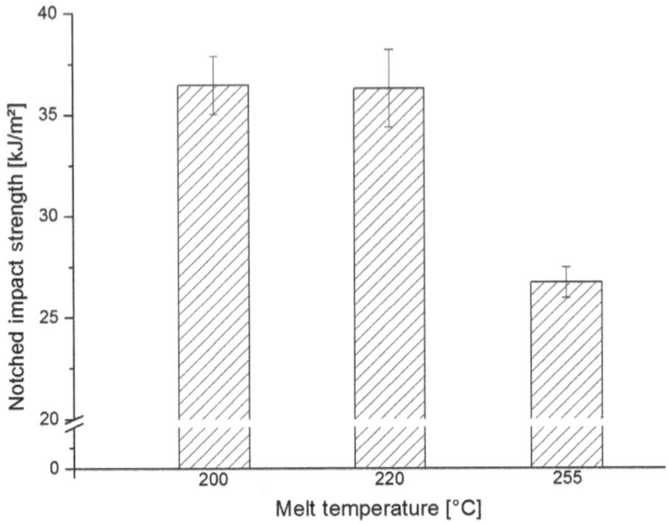

Figure 5.11 Notched impact strength of PP 30Cell depending on the melting temperature in injection molding [20].

The processing temperature of natural fiber-reinforced thermoplasts has a decidedly greater influence on the mechanical properties than that of conventional glass fiber-reinforced thermoplasts. Besides an increased odor development and discoloration of the component (see Figure 5.10), a significant reduction of the energy input during the notched impact experiment results as a consequence of a higher melting temperature (this can be seen in the example of the notched impact strength shown in Figure 5.11). This behavior not only depends upon the melting temperature, but also on the residence time of the natural fiber-reinforced composite melt in the plastification unit of the injection molding machine (not shown in the figure). The reason for the processing-temperature-dependent properties is the thermally sensitive fibers, which degradate significantly at temperatures above 200°C, and therefore, display lower mechanical properties. Accordingly, the material should not be left in the plastification unit for a longer period of time, in order to avoid material degradation, which can occur in correlation with a development of gas. This can induce an exothermic reaction, which, in turn, leads to a heating up of the plastification unit or an igniting of the molding compound.

According to the degradation of biocomposites (e.g., polypropylene with cellulosic fiber), a discoloration induced during the injection molding process can be measured. The *L*-values of the LAB color system tested for samples made of PP with cellulose fibers are shown

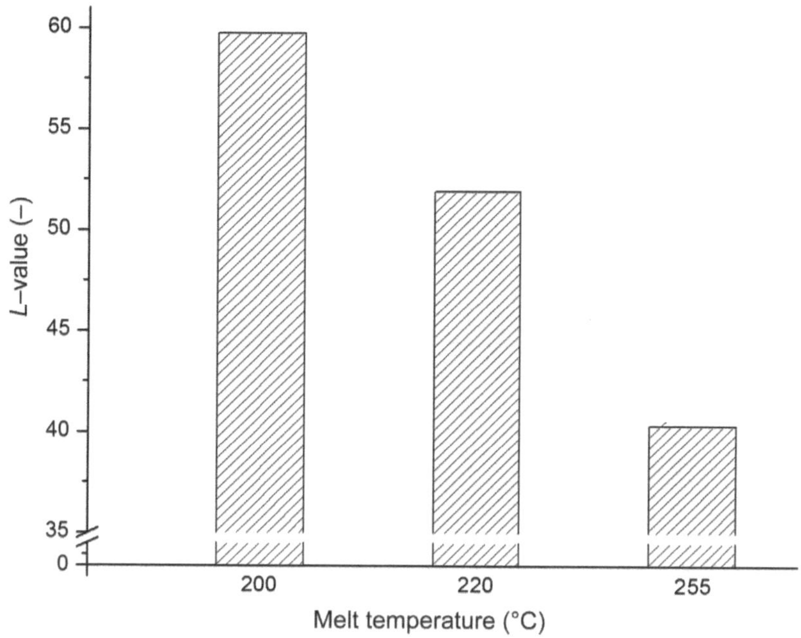

Figure 5.12 Brightness (*L*-value) of PP 30Cell depending on the melting temperature in injection molding [20].

in Figure 5.12. The *L*-value represents the brightness of the sample. A lower value is shown in a darker color. The values in the diagram are allocated to different processing temperatures. The *L*-value decreases as the melting temperature increases. Consequently, a change in color can be identified before the mechanical properties begin to reduce.

Another special feature can be verified in plastics with cellulosic spinning fibers as reinforcement. Owing to the longer fibers in the composite, and the lower stiffness of the cellulose fibers, a complete orientation of the reinforcement fibers in the direction of the flow does not occur (Figure 5.13). Thus, the anisotropy is significantly less defined. Figure 5.14 shows the relative tensile properties of glass fiber and cellulose fiber-reinforced PA 6.10 in correlation with the orientation (0° and 90°). The stiffness of both fiber-reinforced composites vertical to the fiber orientation (90°) equals approximately 50° of the stiffness in the direction of the fiber (0°). In contrast to the relative stiffness of PA 30Cell (70%) diagonal to the fiber orientation (90°), the stiffness of glass fiber-reinforced PAs equals approximately 50%.

Examples of the special structures of bioplastics and bio-based filling materials are shown in the reflection electron microscopy (REM) images of starch blends in Figure 5.15. Depending on which plastic it is, there is

Figure 5.13 Micro CT image taken close to the edge; plan view of sample (width 10 mm) made of PA 6.10 with 30 wt% man-made cellulose fibers [20].

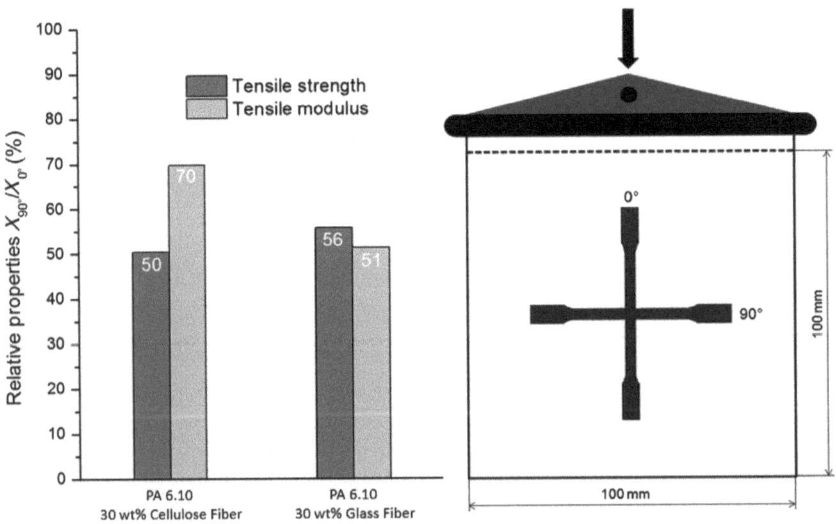

Figure 5.14 Relative properties in the 90° direction (left); schematic illustration of the sample configuration (right); 100% equals the according characteristic in the longitudinal direction [20].

either a better or poorer bonding of the filling material to the matrix. This can be improved by employing a suitable coupling agent. PLA already displays good adhesion without a coupling agent, which can be seen in the separated starch grains in the surface of the fracture.

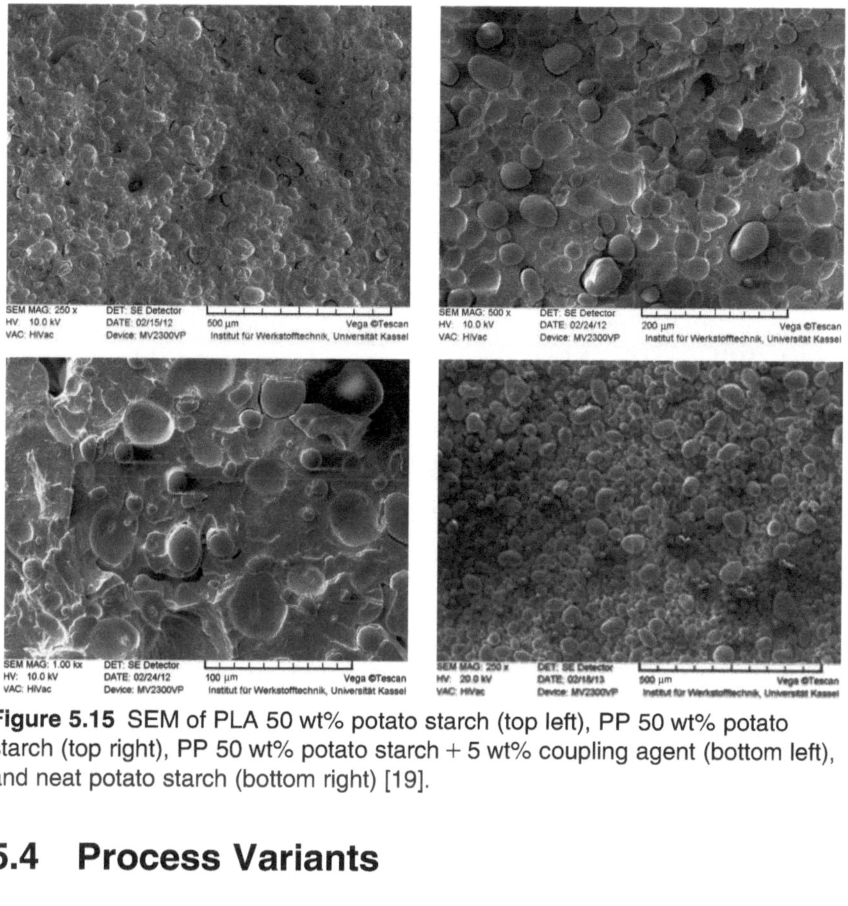

Figure 5.15 SEM of PLA 50 wt% potato starch (top left), PP 50 wt% potato starch (top right), PP 50 wt% potato starch + 5 wt% coupling agent (bottom left), and neat potato starch (bottom right) [19].

5.4 Process Variants

5.4.1 Foam Injection Molding

Natural fiber-reinforced plastics can also be processed like conventional plastics using various foam injection methods. Owing to the hydrophilic nature of the fibers, using water to cause the foaming is especially suitable. In the foam method AquaCell®, water is used as the blowing agent and is incorporated into the melt in the form of very fine natural fibers, which function as carrier particles. Before injection molding, the plastic granulate and the moist natural fibers are blended in a mixer and subsequently inserted into the injection molding machine. One benefit of the AquaCell® method is that, unlike other known physical foaming methods like MuCell®, it does not require an expensive modification of the injection molding machine. Chemical foaming agents are often very expensive, and physical foaming methods require costly investments in machines. In contrast, water is always inexpensive and advantageous from an ecological perspective [44].

In addition to the AquaCell® method, there are further physical and chemical foaming methods available for bio-based plastics. They can also be employed to foam starch blends. However, when chemically foaming PLA starch blends using exothermic and endothermic blowing agents, the degradation products of the blowing agent increasingly lead to an accelerated thermic degradation of the matrix material. This effect first becomes evident in the form of a discoloration of the blend. Moreover, chemically foamed samples display somewhat lower mechanical properties than comparable, physically foamed samples.

Among others, the MuCell® method, developed by the company Trexel (Wilmington, MA) can be used to foam starch blends physically. In this method, the material is melted in the plastification aggregate first and subsequently moves into the gassing area of the plastification aggregate via an additional nonreturn valve. Here, the melt is loaded with gas (N_2 or CO_2). Then the melt containing blowing agent is injected into the cavity of the tool, and foaming of the material occurs due to the pressure drop. The parameters during the loading of the melt with gas play a significant role during the physical foaming of starch blends using the MuCell® method. This can be attributed, above all, to the not-yet-clarified gas absorption of the starch. If the amount of blowing agent is too high, a homogeneous distribution of pores cannot be achieved in the component (Figure 5.16). Analogous to the foamed, conventional plastics, the thickness of the compact outer layer is a decisive influence on the mechanical properties of foamed starch blends.

5.4.2 Injection Molding Compounder

Generally, two steps are used to manufacture natural fiber-reinforced composites. After the first step (compounding), further processing of the manufactured granulate into components is carried out in an injection molding process. An alternative to this two-step process is the injection molding compounder (IMC). The challenge that arises when using this method is the combination of the continuous compounding process with the discontinuous injection molding process. In this method, the compounding is carried out in a corotating twin screw extruder with a screw configuration comparable to that used for the conventional preparation of fiber-reinforced plastics. Using a gravimetric dosing system, the plastic is dosed in a casing of the extruder and

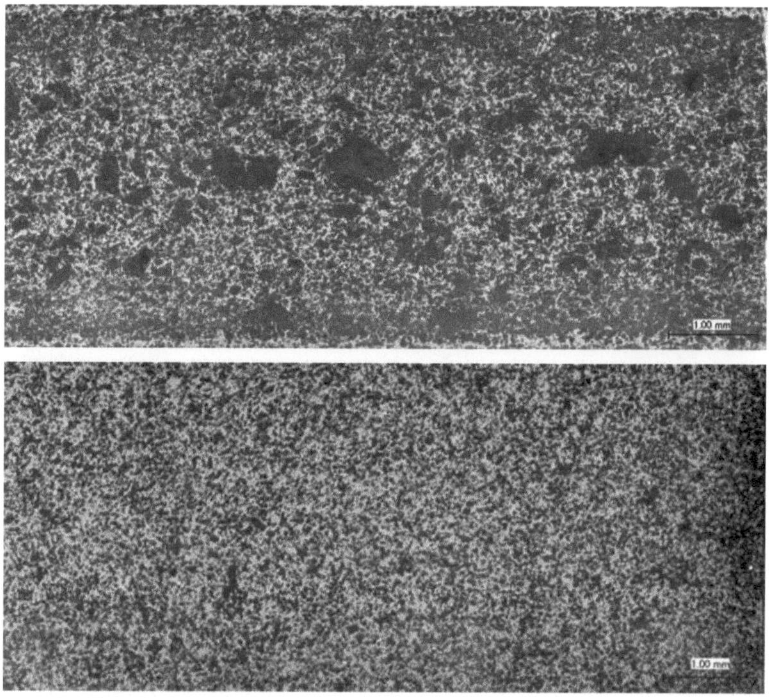

Figure 5.16 Physically foamed starch blend (50 PLA/50 potato starch) with high (top) and low (bottom) gas loading.

melted. Subsequently, the long fibers, which are rolled off of coils, are also dosed via a dosing system in the middle part of the extruder. The plastic and the fibers are then mixed and homogenized.

In a conventional compounding process, the composite would be cooled and made into granules. In contrast, the injection molding compounder continuously feeds the melt into a melt reservoir. From here, the composite continues into the shot-pot, an injection unit that is comparable to a plunger injection molding machine. The processing steps described next are analogous to those of a conventional injection molding process.

Processing long fiber-reinforced plastics using the IMC has several advantages over the usual methods of compounding and injection molding processing (Figure 5.17). The gentle compounding and immediate further processing in the injection molding process minimizes the mechanical shortening of the fibers. Thus, components with essentially longer fibers and better mechanical properties can be realized in comparison to conventional preparation and processing methods. Owing to the one-time exposure of the plastic and the natural fibers to thermal

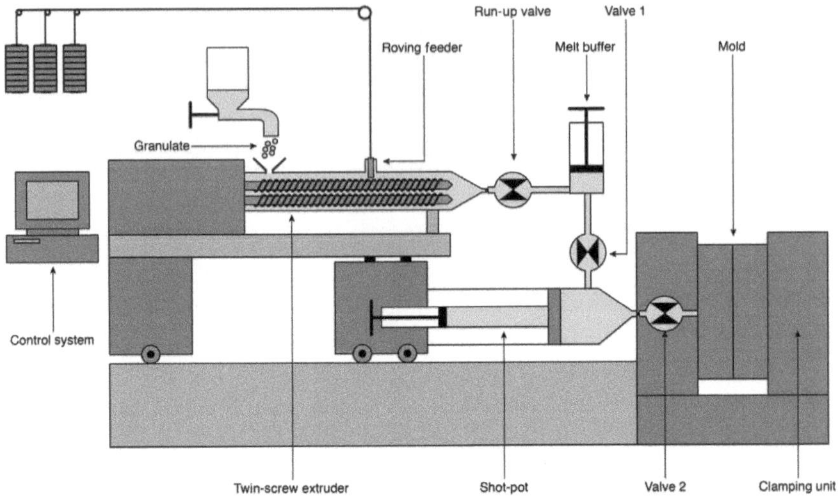

Figure 5.17 Schematic illustration of an IMC according to Refs [45,46].

stress, higher mechanical properties can also be achieved. This applies in particular to the application of temperature-sensitive plastics, like PHB and PLA. Another benefit of IMC is the possibility to alter or adjust the fiber ratio quickly. The components can be directly compared to the specifications, and, if required, the composition can be customized. The high level of throughput, the reduced amount of material used (with no waste), and lower amount of processing energy needed thanks to one-time melting lead to significant cost savings when compared to a conventional process with compounding and subsequent processing in an injection molding process [45–48].

References

[1] Käb H. Biokunststoffe. Kunststoffe 2010;100(10):172–7.
[2] Endres HJ, Siebert-Raths A. 2009. Technische Biopolymere. Rahmenbedingungen, Marktsituation, Herstellung, Aufbau und Eigenschaften, Hrsg.: Hanser Fachbuch, 152.
[3] Stauner, T. 2011. Synthese von hydrophoben Stärkederivaten und Pteroylamidhexylamin-funktionalisiertem Polyvinylalkohol zur Darstellung von Nanopartikeln zum gerichteten Wirkstofftransport. Dissertation, Universität des Saarlandes, 2011.
[4] Fink HP, Ebeling H, Vorwerg W. Technologien der Cellulose-und Stärkeverarbeitung. Chemie Ingenieur Technik 2009;81(11):1757–66.
[5] Avérous L, Halley PJ. Biocomposites based on plasticized starch. Biofuels Bioproducts Biorefining 2009;3(3):329–43.

[6] Li H, Huneault MA. Comparison of sorbitol and glycerol as plasticizers for thermoplastic starch in TPS/PLA blends. J Appl Polym Sci 2011;119 (4):2439–48.
[7] Pushpadass HA, Kumar A, Jackson DS, Wehling RL, Dumais JJ, Hanna MA. Macromolecular changes in extruded starch-films plasticized with glycerol, water and stearic acid. Starch-Stärke 2009;61(5):256–66.
[8] Endres HJ, Kammerstetter H, Hobelsberger M. Plastification behaviour of different native starches. Starch-Stärke 1994;46(12):474–80.
[9] Potente H, Schöppner V, Rücker A. Verarbeitung von Kartoffelstärke auf Kunststoff-Verarbeitungsmaschinen. Starch-Stärke 1991;43(6):231–5.
[10] Janssen LPBM, Moscicki L. Thermoplastic starch: a green material for various industries. Weinheim, Berlin, Darmstadt, Zürich: Wiley-VCH Verlag; 2009.
[11] Röper H, Koch H. The role of starch in biodegradable thermoplastic materials. Starch-Stärke 1990;42(4):123–30.
[12] Zheng X, Chiang WC, Wang SS. Effect of shear energy on size reduction of starch granules in extrusion. Starch-Stärke 1995;47(4):146–51.
[13] Potente H, Heim HP, Ernst W, Föhrer T. Characterisation and modelling of specific enthalpy and heat conductivity of corn grits under consideration of water sorption behaviour. Starch-Stärke 2006;58(2):82–91.
[14] Röper H, Koch H. The role of starch in biodegradable thermoplastic materials. Starch-Stärke 1990;42(4):123–30.
[15] Zheng X, Chiang WC, Wang SS. Effect of shear energy on size reduction of starch granules in extrusion. Starch-Stärke 1995;47(4):146–51.
[16] Schwach E, Averous L. Starch-based biodegradable blends: morphology and interface properties. Polym Int 2004;53(12):2115–24.
[17] Amass W, Amass A, Tighe B. A review of biodegradable polymers: uses, current developments in the synthesis and characterization of biodegradable polyesters, blends of biodegradable polymers and recent advances in biodegradation studies. Polym Int 1998;47(2):89–144.
[18] Wootthikanokkhan J, Wongta N, Sombatsompop N, Kositchaiyong A, Wong-On J, Isarankura na Ayutthaya S, et al. Effect of blending conditions on mechanical, thermal, and rheological properties of plasticized poly (lactic acid)/maleated thermoplastic starch blends. J Appl Polym Sci 2012;124(2):1012–19.
[19] Fuchs J, Feldmann M, Heim H-P. Natürlich und fest anbinden. Kunststoffe 2013;103(8):100–3.
[20] Feldmann M. Biobasierte Polyamide mit Cellulosefasern Verfahren—Struktur-Eigenschaften. Dissertation, Institut für Werkstofftechnik, Universität Kassel; 2013.
[21] Feldmann M, Heim H-P, Bledzki AK. Technische Biokunststoffe mit Naturfaserverstärkung am Beispiel von Polyamid—Cellulosefasern als Alternative zu Glasfasern. Nachwachsende Rohstoffe—Wirtschaftsfaktor Biomasse, C.A.R.M.E.N.—Jahrbuch 2012:2011/2012, pp. 513–20.

[22] Mutlu H, Meier MA. Castor oil as a renewable resource for the chemical industry. Eur J Lipid Sci Technol 2010;112(1):10–30.
[23] Ogunniyi DS. Castor oil: a vital industrial raw material. Bioresour Technol 2006;97(9):1086–91.
[24] Jamshidian M, Tehrany EA, Imran M, Jacquot M, Desobry S. Poly-lactic acid: production, applications, nanocomposites, and release studies. Compr Rev Food Sci Food Safety 2010;9(5):552–71.
[25] Malhotra SV, Kumar V, East A, Jaffe M. Applications of corn-based chemistry. Bridg-Wash-Natl Acad Eng 2007;37(4):17.
[26] Faruk O, Bledzki AK, Fink HP, Sain M. Progress report on natural fiber reinforced composites. Macromol Mater Eng 2014;299(1):9–26.
[27] N. N. DIN 60001: Textile Faserstoffe—Naturfasern Beuth Verlag, Berlin, 1990.
[28] Klemm D, Heublein B, Fink HP, Bohn A. Cellulose: faszinierendes Biopolymer und nachhaltiger Rohstoff. Angew Chem 2005;117(22): 3422–58.
[29] Bledzki AK, Gassan J. Composites reinforced with cellulose based fibres. Prog Polym Sci 1999;24(2):221–74.
[30] Mohanty AK, Misra M, Drzal LT. Surface modifications of natural fibers and performance of the resulting biocomposites: an overview. Compos Interfaces 2001;8(5):313–43.
[31] Mohanty AK, Misra M, Hinrichsen G. Biofibres, biodegradable polymers and biocomposites: an overview. Macromol Mater Eng 2000;276(1):1–24.
[32] Symington MC, Banks WM, West D, Pethrick RA. Tensile testing of cellulose based natural fibers for structural composite applications. J Compos Mater 2009;43(9):1083–107.
[33] Fink HP, Ganster J. Novel thermoplastic composites from commodity polymers and man-made cellulose fibers. Macromolecular symposia. Weinheim: WILEY-VCH Verlag; 2006. Vol. 244, No. 1, pp. 107–118.
[34] Yao F, Wu Q, Lei Y, Guo W, Xu Y. Thermal decomposition kinetics of natural fibers: activation energy with dynamic thermogravimetric analysis. Polym Degrad Stab 2008;93(1):90–8.
[35] Stamm AJ. Thermal degradation of wood and cellulose. Ind Eng Chem 1956;48(3):413–17.
[36] Bilbao R, Millera A, Arauzo J. Thermal decomposition of lignocellulosic materials: influence of the chemical composition. Thermochim Acta 1989;143:149–59.
[37] Tegge G, editor. Stärke und Stärkederivate. Hamburg: Behr's Verlag DE; 2004.
[38] Liu Q, Tarn R, Lynch D, Skjodt NM. Physicochemical properties of dry matter and starch from potatoes grown in Canada. Food Chem 2007;105 (3):897–907.

[39] Tänzer W. 2000. Biologisch abbaubare Polymere: 11 Tabellen; [Polysaccharide, Cellulose, Stärke, Chitin, Pullulan, Polyester, Polyhydroxyfettsäuren, Polymilchsäure, Polycaprolacton, Polyesteramide, Polyphosphophazene, Proteine]. Dt. Verlag für Grundstoffindustrie.

[40] Visser RG, Suurs LC, Bruinenberg PM, Bleeker I, Jacobsen E. Comparison between amylose-free and amylose containing potato starches. Starch-Stärke 1997;49(11):438–43.

[41] Feldmann M, Bledzki AK. Bio-based polyamides reinforced with cellulosic fibres—processing and properties. Compos Sci Technol 2014;100:113–20.

[42] Feldmann M, Bledzki AK, Jaszkiewicz A, Ossadnik S. Method for the production of a fiber-reinforced thermoplastic polymer and extrusion die and device for the temperature control a fiber strand impregnated with a polymer melt. Patent: US 20120199998 A1. 2012.

[43] Harris AM, Lee EC. Improving mechanical performance of injection molded PLA by controlling crystallinity. J Appl Polym Sci 2008;107(4):2246–55.

[44] Faruk O, Bledzki AK, Matuana LM. Microcellular foamed wood-plastic composites by different processes: a review. Macromol Mater Eng 2007;292(2):113–27.

[45] Altstädt V. Inline-Compounding (IMC) zur Verarbeitung hochgefüllter oder langfaserverstärkter Thermoplaste. München: Cluster-Kreis Krauss Maffei; 2007.

[46] Bürkle E, Sieverding M, Mitzler J. Spritzgießverarbeitung von langfaserverstärktem PP. Kunststoffe 2003;93(3):47–50.

[47] Sieverding M, Bürkle E, Zimmet R. IMC-Technik erschließt neue Anwendungsbereiche. Kunststoffe 2005;95(8):34–7.

[48] Klotz B. Injection molding by direct compounding. Specialized molding techniques: application, design, materials and processing. London: Elsevier; 2002. p. 295.

Index

Note: Page numbers followed by "*f*" and "*t*" refer to figures and tables, respectively.